Marine Prawns
and Shrimps

Marine Prawns and Shrimps

Dr. K.P. Biswas
M.Sc., Ph.D. D.F.Sc. (Bom),
E.F (West Germany), F.Z.S, F.A.B.S. (Kolkata)

2011
DAYA PUBLISHING HOUSE®
Delhi - 110 035

© 2011 K.P. BISWAS (b. 1936–)
ISBN 9789351241003

Published by   :   **Daya Publishing House®**
                                 A Division of
                                 Astral International Pvt. Ltd.
                                 – ISO 9001:2008 Certified Company –
                                 4760-61/23, Ansari Road, Darya Ganj,
                                 New Delhi - 110 002
                                 Phone: 23245578, 23244987
                                 Fax: (011) 23260116
                                 e-mail : dayabooks@vsnl.com
                                 website : www.dayabooks.com

Laser Typesetting   :   **Classic Computer Services**
                                 Delhi - 110 035

Printed at   :   **Chawla Offset Printers**
                                 Delhi - 110 052

PRINTED IN INDIA

Dedicated to my Wife
Mrs. Manju Biswas

Preface

Decapod crustaceans, including shrimps, prawns and lobsters are found throughout the world from frigid to tropical zones in both fresh and salt waters, shallow and deep. The density of their distribution is greater in waters lying in the tropical, subtropical and temperate zones in latitudes from 40^0 N to 40^0 S. The Gulf of Mexico, the East and South China Seas and the east coast of Africa abound in these kind of marine products providing vast hauls every year. The majority of crustaceans are marine. They usually possess 20 segments and 19 pairs of appendages. The head and thorax, fused to form what is known as the *Cephalothorax* is encased in a calcareous shell, the carapace. The abdomen is visibly segmented, each segment being dorsally protected by a calcareous surface, like that of *Cephalothorax*. The appendages, the feelers, legs, swimmerets etc are modified to suit the various functions to which they are adapted. Respiration is carried on by the branchia or gills which are enclosed in carapace. Their sexes are seperate. Prawns carrying fertilized eggs are sometimes referred to as "Berried" prawns. Between the egg and adult stages, there is frequently a striking, varied and interesting metamorphosis.

All prawns have hard shells on the outside, which affords protection to the animal and supports the internal organs. The shell also provides points of attachment for the muscles with which the animal moves. It plays, therefore, the part of a skeleton, but unlike ours, the exo-skeleton of the prawn is outside the body. As it does not increase in size after it has been formed and is inflexible, the animal has to cast its shell periodically as it grows. During moulting, the shell splits and is cast away leaving the animal with only a thin flexible membrane. This, however, rapidly grows by the absorption of water and gradually hardens by the deposition of calcium salts. The process goes on over and over again. Some of these animals have the power of voluntarily throwing off their limbs and regenerating them later.

All species of prawns are edible. They have the property of storing great quantities of glycogen or animal starch and of fat, which renders them nutritious. They are rapid growing and prolific animals, capable of extensive fishing.

Prawns are generally of a very timid and retiring disposition, hiding themselves in holes and crevices of the surface on which they live. At the slightest approach of danger, they hurry away to their burrows or lie quiescent on the substratum where in the midst of weeds etc they generally escape unnoticed. Prawn is produced in the tropics for consumption across the world. Prawn is not a staple protein, but a luxurious item and will not be bought when the family is on a budget.

Shrimp still has the image of a luxury item on the restaurant trade. Marine prawns, tastier and costlier than their freshwater counter parts, have become an avenue to earn foreign exchange by the developing countries of South East Asia, since mid 1950's. They catch these prawns from inshore and offshore coastal waters and export to European countries and U.S.A at a high price.

The marine prawn fishery, which supports the export industry is generally confined to shallow coastal areas within 40 m depth and comprised of the species like, *Penaeus inducs, P. merguiensis, P. monodon, P. semisulcatus, Metapenaeus dobsoni, M. affinis, M. monoceros, M. brevicornis* and *Parapenaeopsis stylifera*. Though different species show variations in the breeding season, most of them spawn throughout the year, having two peak periods, November to December and February to April, the former being most productive.

From the catch analysis of different centres, it has been found that the growth of various prawns differs according to the species and locality and most of them live for about 2 years. But the growth is rapid in early stages and slows down after its maturity. The normal growth rate is 2.5 to 5 mm per month. The maximum growth of 25 mm per month has been recorded in *Penaeus monodon*. An almost similar growth rate of 24.2 mm per month has been observed in *P. indicus* at Chennai.

Large quantities of shrimp are produced in developing countries for export to Japan, United States and Western Europe; and infact shrimp trade has been a multibillion dollar industry. Shrimp is one of the major seafood export product from India and the country is the fourth largest exporter of shrimp in Asia with a 2.43 per cent share in the 50 billion dollars world seafood market. Share of shrimp in India's total marine product export was 54 per cent in 2006-07, primarily to Japan, USA and European Union. (EU).

The rapid changes in the value of dollar against the Indian currency have been negatively affecting the Indian shrimp exports. The increasing competition by other shrimp producing countries (Bangladesh, Vietnam and Thailand) that harvest *Vannamei* shrimp *(Penaeus vannamei)* with lower production cost and shorter duration of culture period compared to Indian black tiger shrimp has also negatively affected the shrimp trade. The share of Indian shrimp at USA shrimp market which was as high as 11.2 per cent during 2002 declined to 3.96 per cent during 2006. Over a span of 6 years, (from 2001 to 2006) Indian shrimp exports to Japan market has been reduced by 30 per cent . The major species of shrimp exported to Japan were Black tiger prawn *(Penaeus monodon)*, white prawn *(Penaeus indicus)*, Brown *(Penaeus aztecus)* and Flower *(Penaeus semisulcatus)*. Initially, quality problems, especially the muddy mouldy smell emerging from cultured shrimp led to decrease in Japanese imports of black tiger shrimp from India. A comparison of the unit value realized by Indian shrimp exports to Japan with shrimp exports from competing countries revealed that India (1.06-0.78) realized lesser unit value than Sri Lanka (1.34-1.16), Thailand (1.22-0.9), Indonesia (1.22-1.05) and Vietnam (1.02-0.98).

The main competition that India faces with regard to shrimp trade comes from Vietnam and Thailand. Thailands shrimp production by culture during 2004 was 3,25,000 tonnes as against India's production of 143170 tonnes in 2005-06. India's advantage is that it can increase its production by extending the farming area. The production costs in India are also lower. In the Japanese market, the demand for processed shrimp varieties such as tray pack, sushi, and bread-battered items are gradually increasing. Taking the advantage of the growing demand for these products, Indian exporters should take up value addition of products. Diversification of markets has to be explored to bring the Indian shrimp trade to benefit.

Since third quarter of 2008, there have been reports of low demand of prawns. Thailand's export have declined (145983 tonnes in first half of 2008 in comparison of 348488 tonnes in 2007). Indonesian exports mainly to the US and Japan totalled 157544 tonnes, but for the first nine months in 2008, only 120000 tonnes were exported.

Consumers now want guarantees of food safety and full traceability. The product require certification for environment friendly and antibiotic free. The implementation of "farm to fry pan" systems, quality control and environmental protection has been taken up in most countries. Thailand, the leading Asian producer, have 172 registered prawn farms and 18,365 GAP certified farms producing 95 per cent of prawn from aquaculture, enabling full control of quality and safety. They are aggressively promoting the brand "Thaifarm prawns" to gain market recognition.

Dr. K. P. Biswas

Contents

Contents

Introduction

The term 'prawn' is identical with the term 'shrimp' used in western countries. At the Prawn Symposium of the Indo-Pacific Fisheries Council held at Tokyo in 1955, the decision arrived at was that the term 'prawn' should be applied to Penaeids, Pandalids and Palemonids and the 'shrimp' to the smaller species belonging to other families.

Marine prawns belonging to class crustacea are benthic living on the sea bed even upto a depth of 8000 metres.

The body of a crustacean is segmented and covered by a chitinous integument, the exoskeleton. Typically each segment has a ring of exoskeletal covering, the adjacent rings being connected by thinner cuticle, the arthroidal membrane, making feasible the movement. The dorsal region of the ring is called, tergite, and the ventral, the sternite. A pair of appendages are found on the lateral sides of the sternite portion when tergite overhang freely over the sternites, they are called pleurons.

The outer integument of a crustacean is made up of several layers. On the posterior part of the cephalic region a dorsal shield or carapace arises as an integumental fold, which may take the form of a bivalve shell, a shield or mantle in the lower crustaceans. In the prawn family the carapace fuses with some or all the tergites of other thoracic segments, the posterior extremity projecting freely at the sides to form branchial chambers. The anterior region may be produced into a toothed prolongation, the rostrum.

The appendages or limbs of prawn are of three types, the uniramous, biramous and multiramous. A limb has a basal portion, which is attached to the body, the protopodite consisting of two segments, the proximal coxa and the distal basis. A precoxa also may be present proximal to the coxa. In a uniramous limb only one ramus develops on the protopodite, as in the antennules of nauplius larvae and also of several adult crustaceans. If two rami develop on the protopodite the limb is biramous, the inner one is the endopodite and the outer one the exopodite. The endopodite is usually made up of five segments or podomeres. From the proximal to the distal extremity the podomeres are the ischium, merus, carpus, propodus and dactylus respectively. Most of the appendages other than antennae in the higher crustaceans are biramous, but due to suppression of one ramus some limbs may appear as uniramous.

On the protopodite leaf-like processes may develop on the outer margin called epipodites or exites, and on the inner margin, the endites. A foliaceous limb may have a protopodite or axis attached to the body, which in some cases is made up of a few segments or podomeres. On the inner margin of the axis there are generally six endites, the proximal one directed inwards in the axial direction of the body. On the outer margin there are two exites, the proximal one being called bract and the distal one, the flabellum. The structures are mostly flattened and leaf-like. Due to the action of the appendages food particles are directed forward to mouth along the mid-ventral groove of body, the proximal endites helping in masticating the food and act as gnathobases. All or some of the epipodites may be respiratory.

The marine prawn fishery of India remained largely in the domain of the traditional fishing sector till about the middle of 19th century, with a low production. The discovery of rich prawn grounds in offshore waters and the introduction of mechanized fishing especially by shrimp trawlers, together with modern processing techniques and increased demand for processed prawns in the international market, together contributed very rapid development in capture fisheries of marine prawn. Today, prawns are intensively exploited throughout the Indian coast upto about 100 metres depth by a large variety of fishing enterprises, including small, medium and large trawlers and the mechanized and non-mechanized traditional crafts. While the normal fishing ground of small trawlers lie within the 50 m depth contour in most parts of Indian coast, the large trawlers of 20-25 m OAL operating in the north-eastern region cover deeper areas of continental shelf upto 100 m.

Due to the declining trend of prawn catch from the conventional fishing grounds in recent years, many of the small and medium size trawlers, engaged in coastal fishing have extended their range of operation to deeper waters to increase the catch returns. Long trips involving 2 to 5 days fishing voyage have been resorted to in large scale at many parts of the coast, particularly in Kerala, Karnataka and Saurashtra. In the traditional sectors new gears have been introduced or existing gears modified to catch more prawns. Widespread introduction of ring seines (a smaller version of purse seine) along the coasts of Kerala and Karnataka and that of minitrawls along the coast of Kerala, which have proved to be very efficient in catching prawn and fish were major improvement since mid-eighties. The bottom-set gill nets commonly used for catching large-sized prawns have been replaced by a more efficient selective 'trammel' net. This triple walled entangling net, locally known as 'Disco net' have soon became very popular for its impressive performance; and is now one of the most common gears employed in the traditional sector along the south west coast and throughout the east coast of India.

Prawn fishing is done almost throughout the year on most part of Indian coasts except for a short period of off-season during monsoon period. In Kerala, however, peak fishing takes place during the south-west monsoon period by trawlers.

In the total marine production of India, prawns constitute 10-12 per cent, the major portion, about 80-85 per cent is contributed by the west coast. The prawn landings of 85 thousand tonnes in 1958 increased to 389 thousand tonnes in 1998 showing a four fold increase over the years.

Considerable variations exist in the concentration of prawn resources in different regions of Indian coast, which may be attributed to the differences in the extent of continental shelf, bottom conditions, hydrographic features, available estuarine areas and climatic conditions, which greatly influence the distribution and abundance of prawn stocks. About 80 per cent of the prawn catch is obtained from Maharashtra, Gujrat and Kerala alone. The other maritime states like Tamilnadu, Andhra Pradesh and Karnataka together account for about 15 per cent and rest by other maritime states and Union Territories.

The traditional fishing sector landed about 49 per cent and the shrimp trawlers 51 per cent during 1978. While Kerala, Maharashtra, Gujrat and Tamilnadu inorder of abundance landed major portion of prawn from trawl catches, Gujarat and Maharashtra accounted bulk of fixed bag net (dol net) catches, Kerala, Tamilnadu and Andhra Pradesh dominated in the boat seine and shore seine catches, Kerala in ring seine catches and Kerala and Andhra Pradesh in gill net catches in traditional sector.

Indian prawn fishery is multi-species in character and supported by several species of penaeid and non-penaeid groups. From export point of view, the penaeid prawns are more important than the non-penaeid prawns on account of their larger sizes and high unit value. The penaeid prawn on an average is of about 65 per cent and non-penaeid prawn 35 per cent in the total fishery. While penaeid prawns constitute almost the entire catch in Kerala, Tamilnadu, Karnataka, Orissa, Goa, and Pondicherry, the non-penaeid prawns accounts for as much as 65 per cent in Gujrat, 38 per cent in Maharashtra, 45 per cent in West Bengal and 12 per cent in Andhra Pradesh besides the penaeid prawn in the fishery.

Penaeid prawns support the trawl fishery mainly (92 per cent). The dominant species forming the trawl catches are *Solenocera crassicornis, Metapenaeus monoceros, Parapenaeopsis stylifera, P. hardwickii,* along the Gujrat coast, *M. affinis, M. monoceros, P. stylifera, P. hardwickii* and *P. sculptilis* along the Maharashtra coast. *M. monoceros, M. affinis, P. merguiensis* and *M. dobsoni* along the Goa and North Karnataka coast. *P. stylifera, M. dobsoni, P. indicus, M. monoceros* and *M. affinis* along the southwests coast. *P. indicus, P. semisulcatus* and *M. dobsoni* along the south east coast, *M. monoceros, M. dobsoni, P. stylifera, M. brevicornis, P. indicus, P. monodon* and *M. affinis* along the Andhra coast and *M. monoceros, P. indicus, P. monodon, P. merguiensis* and *P. hardwickii* along the north east coast. The traditional fishery is constituted by both the penaeid and non-penaeid prawns.

The fixed bag nets operating in the inshore waters of Gujrat and Maharashtra land the bulk of non-penaeid prawns (70-75 per cent) in the country of which three-fourth is constituted by *Acetes* spp. Among the penaeid prawns, caught in the indegenous gears, the most important are *Metapenaeus kutchensis, M. dobsoni, P. stylifera* and *P. indicus* in the southwest coast and *P. indicus, M. dobsoni, M. monoceros, M. brevicornis* and *P. monodon* along the east coast. Among non-penaeids, besides *Acetes* spp, *Nematopalaemon tenuipes* and *Exhippolysmata ensirostris* contribute substantially to the dol nets catch in Bombay-Saurashtra coasts and *Exopalaemon styliferus* in West Bengal coast.

The recent shift in night and offshore trawling resulted in the occurrence of a number of non-conventional prawn species throughout the west coast and southeast coast. In Karnataka and Kerala the catch of *M. monoceros* has increase considerably in the trawl landings as the trawlers have started operating outside 50 metre depth line, where the concentration of the species is more. Night fishing conducted by the trawlers in voyage trips has increased the catch of the species, as it is nocturnal in habit. Similar increase in the catch of *M. monoceros* and other associated species like *P. canaliculatus, P. semisulcatus Metapenaeopsis stridulans, Trachypenaeus* spp. has been observed in the night catch of shrimp trawlers operating in Kerala coast. In Bombay waters also, the catches of non-conventional species like, *M. stridulans, Trachypenaeus curvirostris, Parapenaeus longipes* and *Solenocera choprai* have increased as a result of fishing in deeper waters. In Tamilnadu coast also a number of non-conventional species have made their appearance in sizable quantities in the trawl fishery due to extension of fishing into deeper waters.

Estuaries and backwaters serve as nursery grounds of many species of penaeid prawns and form an important source of recruitment for the inshore stocks. The Hoogly-Matlah estuarine system, Chilka lake, Godavary estuary, Kakinada backwaters, Pulicat lake, Ashtamudi lake, Cochin backwaters and the Rann of Kutch support a lucrative fisheries for prawn juveniles like, *P. indicus, P. monodon, M. dobsoni, M. kutchensis,* and *M. monoceros.* About 20-25 thousand tonnes of juvenile prawns are exploited every year from the brackish water systems in the country.

The latest potential stock of marine prawns was calculated to be 232 thousand tonnes for the inshore areas upto 50 m depth and 4 thousand tonnes between 50 and 100 m depth on the North-East coast. The annual catch data indicate that the production of marine prawns has exceeded far beyond the catchable potential of 236 thousand tonnes in 1990. Sock assessment of important prawn species like *M. dobsoni, M. monoceros, P. stylifera, P. indicus, P. semisulcatus, P. monodon* and *A. indicus* has revealed that the coastal prawn resources of India have been fully exploited. Even over fishing has been observed in respect of species like *M. dobsoni, P. indicus* and *P. semisulcatus*. The change in fishing pattern of shrimp trawlers in recent years such as, venturing into the high seas leaving the conventional grounds, voyage fishing, involving night trawling throughout the west and southeast coast, the successive increase in production of marine prawns are maintained.

The extent of productive areas of deep-sea prawns was estimated at 5000 square kilometer and the potential resources seems to be about 5300 tonnes/year. The highest density of prawn (nearly 40 per cent) have been observed between 275 and 375 m depth (high catch rates of 113-224 kg/hr). The catches were composed of about a dozen of new species belonging to both penaeid and non-penaeid groups of which six species namely, *Heterocarpus woodmasoni*, (41 per cent), *H. gibbosus* (7 per cent), *Parapandalus spinipes* (22 per cent), *Plesionika martia* (4.5 per cent), *Penaeopsis rectacuta* (9 per cent) and *Aristeus semidentatus* (10 per cent) formed the major components. In the Gulf of Mannar, the resource was found to be less abundant than in south west coast and the catches predominantly represented by *H. gibbosus, A. semidentatus, Solenocera hextii* and *Metapenaeopsis* spp. Commercial fishing for this new resources has started in recent years along the Kerala coast.

The key species of penaeid prawns in the recent emergence of culture-based fisheries take advantage of the natural ecosystem of the sea. Stocking natural waters with prawn fry is to restore the once damaged fry resources of coastal waters or to stimulate the potential productivity of the area by the fry set free artificially.

The aquaculture of prawn is a multi billion-dollar industry attracting many developing countries including India for local employment, production of protein food and foreign currency. Prawn farming has been practiced for more than a century for food and livelihood of coastal people in many Asian countries, like Indonesia, Philippines, Taiwan, China and India. From 1970-1975, research on breeding was conducted and monoculture techniques in small ponds were developed at the Tangkang Marine laboratory in Taiwan provice of China and at the centre Oceanologique du Pacifique in Tahiti in the south Pacific. In Thailand extensive and semi-extensive farms were commercially established in 1972 and 1974 respectively, after the first successes in breeding of *P. monodon* at Phuket Fisheries Station in 1972. Between 1980 and 1987 there was a boom of small-scale intensive farms in Taiwan provice of China due to commercial success in formulated feed development, mainly to produce shrimp for export to Japan. A viral disease outbreak caused the collapse of the industry in Taiwan provice of Chaina in 1987-1988. This led Thailand, encouraged by extremely high price in the Japanese market due to short supply, to replace the Taiwan provice of China as the world's leading producer of farm raised *P. monodon* in 1988. Later, the culture of this species spread throughout the southeast and south Asia, as it can grow upto a large size (40-60 gram) with high value and demand in international market. The introduction or importance of wild broodstock is commonly practiced among the major producing countries because local supplies are insufficient and demonstration technology has not yet been commercially developed. However, disease free broodstock are highly desirable.

Frozen head on, head off and peeled shrimp used to be the major products for export to the main markets, which are USA, Europe and Japan, Financially *Penaeus monodon* is the most important traded aquaculture commodity in Asia. Cost and Freight (C and F) prices in Japan, whose market mainly requires large headless prawn from extensive and semi-intensive farms in India, Indonesia and Vietnam varied from US $ 9-14 /kg during 2001-2004. The U.S. market purchased mainly small headless shrimp from intensive farms of Thailand and India at Cost and Freight (C and F) price ranging from US $ 7-13/kg during the same period. The Europe market, which mainly requires small head-on shrimp from South-East Asia intensive farms, paid C and F prices between US $ 4.7 and 9.0 /kg during 2001-2004. Total aquaculture production of *P. monodon* increased from 21.000 tonnes in 1981 to 200,000 tonnes in 1988. It then sharply increased to nearly 500000 tonnes with a value of US $ 3.2 billion in 1993. Since then, production has been quite variable, ranging from a minimum of 480000 tonnes in 1997 to a maximum of 676000 tonnes in 2001.

Among the prawn culture industry, *Penaeus monodon* is one of the most valued species, which has contributed about 75 per cent of the total aquaculture production. However, the prawn industry has been threatened by the outbreak of various infectious diseases, mainly of viral origin, which caused a decrease in prawn production in India.

Prawn farming *(P. monodon)* has been threatened mainly by viral and very few bacterial diseases. Luminous bacteria and *Vibrio* species are the major causative agent for bacterial disease of prawn, but these are not so significant in comparison to viral diseases.

There are five major viruses that create problems of *P. monodon* aquaculture. They are white spot syndrome virus (WSSV), yellow head virus (YHV), hepatopancreatic parvo like virus (HPV), infectious hypodermal and hematopoietic necrosis virus (IHHNV) and monodon baculovirus (MBV).

White spot disease caused by WSSV is regarded to be the most serious problem faced by the shrimp aquaculture industry all over the world. In Ecuador, viral disease, particularly Taura and white spot syndrome virus decreased the production by 58 per cent in 1994. The virus was first reported to be found in *Penaeus japonicus* in Taiwan in 1992.

WSSV infects most of the prawn tissues like sub cuticular epithelium, gills, lymphoid organs, antennal gland, hematopoietic tissues, connective tissues, ovary and the ventral nerve cord. The main clinical symptom is the presence of white spot ranging from 0.5 to 2.0 mm in diameter on the exoskeleton and epidermis of the diseased prawn. The other signs are rapid reduction in food consumption, lethary, loss of cuticle and often-generalized reddish to pink discoloration.

Infection caused by WSSV can reduce production vigorously and it infects from post-larval to adult prawn. WSSV is a very contagious virus and the infection spreads both vertically and horizontally.

There are numerous diagnostic methods available for detection of WSSV like microscopy, histopathology and reverse passive latex agglutination assay. The most sensitive methods are rapid molecular methods, such as, gene probe and polymerase chain reaction.

Chapter 1

Identification and Distribution of Marine Prawns

Marine prawns are generally confined to shallow coastal waters, within 40 metre depth. They are of different sizes and only those which are of large size and available in plenty are exploited for commercial purposes. Among the marine prawns, some species form good fishery, some are important as occasional small fishery and still others, mostly deep sea dwellers have recently been discovered to form potential fishery. Several species are caught in very small quantities, but they also add to the total catches.

The important characters considered in the identification of prawns are the carapace and its spines, the rostrum and its dorsal and ventral teeth, the ridges or carinae, the grooves or sulci, abdominal carination, telson, appendages and their segments, petasma, and appendix masculina in the male and the thelycum in the female.

The body of the penaeid prawn is divided in two segments, the cephalothorax and abdomen. Most organs, such as gills, heart, hepatopancreas and stomach are located in the cephalothorax, while the gut and the reproductive organs are located in the muscled abdomen. The appendages of the cephalothorax are modified into different forms, including five pairs of walking legs (pereiopods), jaw-like structures, antennule and antennae. On the abdomen, five pairs of swimming legs (pleopods) are located. The heart and the main haemolymph vessels are located dorsally, together with the haemocoel. To enable growth the prawn has to moult periodically (loose the extracellular cuticle from the underlying epidermal layer to replace it for a new flexible, expandable exoskeleton that subsequently hardens).

Penaeid prawn belonging to the sub class Malacostraca, order Decapoda, has nineteen pairs of appendages, five cephalic, eight thoracic and six abdominal. The head appendages are the antennules (1st antennae), antennae (second antennae), mandibles, maxillulae (1st maxillae) and second maxillae. The first three thoracic appendages are modified as maxillipeds and directed forward for masticatory purposes, and the rest five are walking legs or pereiopods. The abdominal appendages are the pleopods or swimming legs. The antennules are uniramous and the rest are biramous.

Antennules are the anteriormost limbs situated below the level of eye-stalks. The protopodite has three podomeres, the proximal precoxa, middle coxa and the distal basis, and on the latter are situated two short many-jointed flagellae. The precoxa has a hollow depression on one side into which eye can be fitted in. In this very segment statocyst is situated with its opening at the base.

Antennae are situated posterior to the antennule. The protopodite has two segments, the coxa and basis, on which an exopodite and endopodite are situated. The exopodite is flat and broad as scale-like, called the squame. The endopodite is very long with three basal podomeres and a narrow and many jointed flagellum, about one and half times the length of the whole body. The ducts of excretory canals open on the coxae off the antennae.

Mandibles are present on either side of mouth. The protopodite of each is modified into a stout calcified structure with toothed inner edge, and the edges of the two on either side act to grind food to size. The endopodite is located on this as a palp, made up of two joints on whose edges short hairs are present giving a plumose appearance. The exopodite is absent.

Maxillulae are very small appendages in comparison to others. The protopodite has two leaf-like processes, the coxa and basis with flattened edges bearing sharp hairs, the gnathobases, for helping in

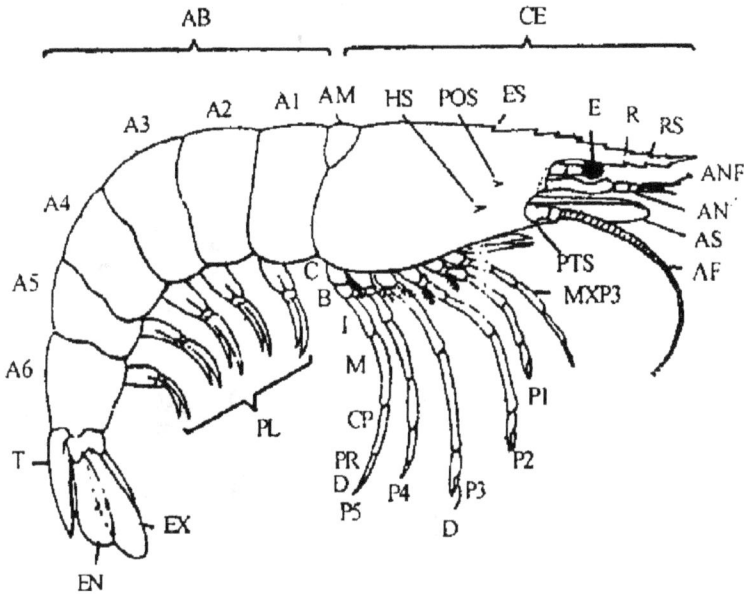

Figure 1: Diagrammatic View of a Penaeid Prawn Showing Taxonomic Characters

AB: Abdomen; AF: Antennal flagellum; AM: Arthroidal membrane; AN: Antennule; ANF: Antennular flagellum; AS: Antennal scale; A1–A6: First to sixth abdominal segments; B: Basis; C: Coxa; CE: Cephalothorax; CP: Carpus; D: Dactylus; E: Compound eye; EN: Endopodite; ES: Epigastric spine; EX: Exopodite; HS: Hepatic spine; I. Ischium; M: Merus; MXP3: Third maxilliped; PL: Pleopods; POS: Postorbital spine; PR: Propodus; PTS: Pterygo-stomian spine; P1–P5: First to fifth pereopods; R: Rostrum; RS: Rostral spine; T: Telson.

masticating food. The endopodite is also leaf-like and unjointed, the distal end getting narrower, and the edges covered with small hairs. The exopodite is absent.

The protopodite is flat in maxillae and produced into four leaf-like lobes, edges being flattened with pointed hairs for masticatory purposes. The exopodite is also flattened and very large, expanded and boat-shaped, called the scaphognathite. Its action produces currents for water to pass through the gills chamber. The endopodite

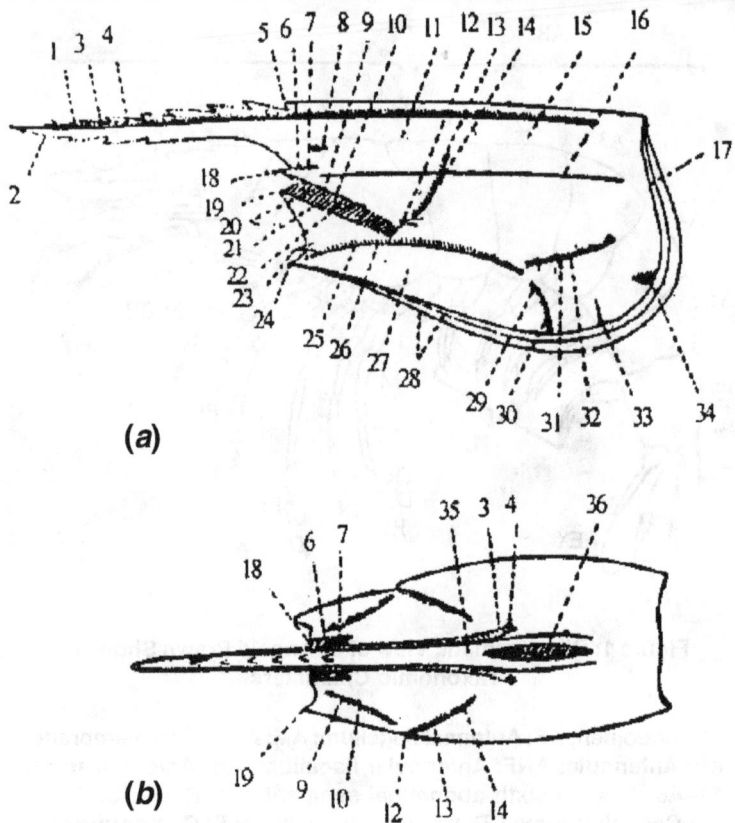

Figure 2: (*a*) Side View of the Carapace; (*b*) View from Above of the Carapace, of a Penaeid Prawn Showing Taxonomic Characters

1: Rostral tooth; 2: Ventral rostral tooth; 3: Adrostral carina; 4: Adrostral sulcus; 5: Epigastric spine; 6: Gastrofrontal carina; 7: Gastrofrontal sulcus; 8: Postocular sulcus; 9: Orbito-antennat sulcus; 10: Gastroorbital carina; 11: Gastric region; 12: Hepatic spine; 13: Cervical sulcus; 14: Cervical carina; 15. Cardiac region; 16: Longitudinal suture; 17: Pterygostomian sulcus; 18: Orbital or supraorbital spine; 19: Antennal spine; 20: Postorbital spine; 21: Antennal carina; 22: Postantennal spine; 23: Branch iostegal spine; 24: Pterygostomian spine; 25: Hepatic carina; 26: Hepatic sulcus; 27: Pterygostomian region; 28: marginal region; 29: Inferior carina; 30: Inferior sulcus; 31: Branchiocardinal carina; 32: Branchiocardinal sulcus; 33: Transverse suture; 34: Stridulating organ; 35: Postrostral carina; 36: Postrostral or median sulcus.

is unjointed, small and leaf-like situated between the protopodite and exopodite.

The first maxilliped has a foliaceous appearance. The protopodite is split into two proximal small lobes and a distal bigger one, both provided with hairs on their edges. The exopodite is also flattened and leaf-like with a broader base and narrower end, edges fringed with hairs. The endopodite is narrow and made up of a number of segments. Proximal to the protopodite and exopodite there is a flattened triangular structure, the epipodite, which has a respiratory function.

The protopodite of second maxilliped is of the normal type, with a coxa and basis on which stand the exopodite and endopodite. The latter is made up of five segments, typical of the endopodites of all thoracic segments, the ischium, merus, carpus, propodus and dactylus. The endopodite is curved distally giving the shape of an interrogation mark. The exopodite is slightly flattened with striations and setose hairs on the margin. A Y-shaped flattened epipodite is attached to the stem of coxa. This helps in respiration.

The third maxilliped is similar to the second maxilliped, but the endopodite is straight with five segments. The exopodite is slightly flattened with striations and hairy edges. On the coxa there is a Y-shaped respiratory epipodite.

All the first three pereiopods are identical in structure except in size, the third one being the longest. The protopodite of each has two segments, the coxa and basis. The endopodite is five-segmented but the dactylus of each limb is hinged to the sides of propodus articulating against it. This is known as chelate articulation and the leg is called cheliped, used for holding food particles and taking them to the mouth. The exopodites of these are minute and fringed with hairs. Y-shaped epipodites, respiratory in function, are attached to the coxa of each leg. These legs are also used for walking.

The size of fourth and fifth pereiopods vary, the fourth being longer than the fifth. These limbs are also of the same pattern as the first three, but have non-chelate endopodites. Exopodite of fourth is minute and fringed with hairs. The fifth leg has a still smaller exopodite, which is a pointed transluscent process without hairs. Epipodites are absent on these legs. These legs are also used for walking.

The first pleopod differ in male and female. In both the protopodite is made up of a coxa and longer basis, and the exopodite and endopodite are flattened and thick, fringed with hairs. Endopodite is absent in the female or in some cases it may be present as a very small bud-like process. In the male endopodites are provided with hooks and those of the two sides are interlocked to form a rod like structure, the petasama through which sperm is transferred into the thelycum of the female.

Second of fifth pleopods are all similar conforming to the typical biramous pattern in both sexes. In males there is a process, appendix masculina, attached to the endopodite of the second pleopod. The exopodites are bigger than endopodites.

The sixth abdominal limb is the uropod which is biramous, constructed on the same pattern as other pleopods, but bigger in size. The protopodite is of a single segment due to the fusion of a coxa and basis. The two uropods together with the telson form a tail fin. This acts as a balancing organ and is also used by sudden flexion for leaping backward.

Hippolysmata ensirostris, characterized by longer rostrum than carapace, with an elevated dentate basal crest with acute and unarmed apex and concave lateral margin of telson, is distributed in India, Sri Lanka, Mayanmar and Sumatra. In India the species is found along the coasts. They attain a maximum size of 80 mm and form a fairly good fishery in Mumbai and Godavari estuary. It also form small fishery in other areas.

Hippolysmata vittata occur in South Africa through Red Sea, Persian Gulf and India to East Indies and Japan. The species is recognized by shorter rostrum than carapace and without elevated basal crest. Lateral margin of its telson is convex. The apex is blunt with a pair of spines. The species is found only in small numbers and attain length of 40 mm.

Dactylus of last three pereiopods of *Palaemon styliferus* are not abnormal in length, that of third scarcely one and half length of propodus and that of fifth at most one third length of propodus. Pleopods are normal in length. One or more subapical dorsal teeth or rostrum. Last four abdominal somites bluntly carinate dorsally. The species is distributed in West Pakistan and India to Malayan Archipelago in shallow coastal waters, brackish and sometimes

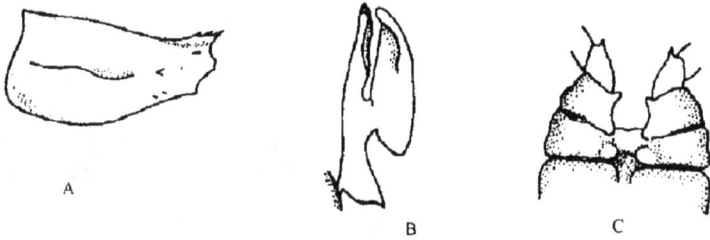

Figure 3: (A) Carapace; (B) Petasma and (C) Thelycum of *Acetes indicus*

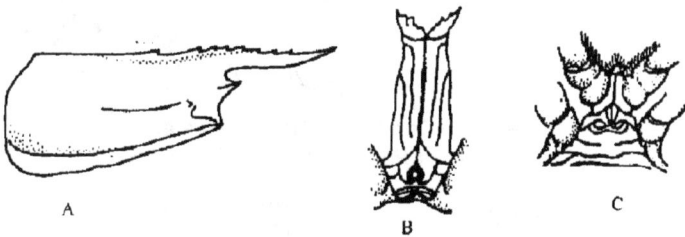

Figure 4: (A) Carapace; (B) Petasma and (C) Thelycum of *Metapenaeus affinis*

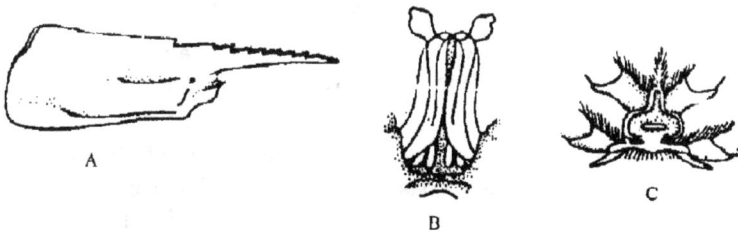

Figure 5: (A) Carapace; (B) Petasma and (C) Thelycum of *Metapenaeus monoceros*

freshwater areas. In India the species occur in the northern region of both coasts. It attain a maximum size of 90 mm and form one of the most important fisheries in Gangetic Delta.

Dactylus of last three pereiopods of *Palaemon tenuipes* are very long and slender. The fourth and fifth pairs are excessively long, flagelliform with dactylus much longer than carapace. Pleopods are very long, first pair much longer than carapace. Carpus of second pereiopod much more than half as long as palm, Basal crest of

Figure 6: (A) Carapace; (B) Petasma and (C) Thelycum of *Metapenaeus dobsoni*

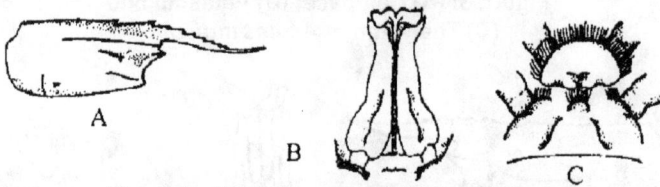

Figure 7: (A) Carapace; (B) Petasma and (C) Thelycum of *Parapenaeopsis hardwicki*

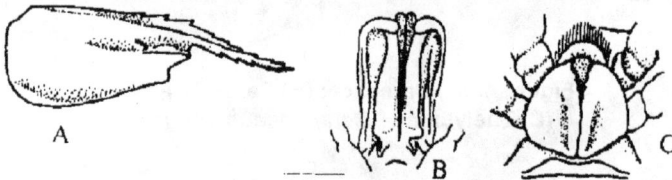

Figure 8: (A) Carapace; (B) Petasma and (C) Thelycum of *Penaeus merguiensis*

rostrum with atmost seven teeth. The species occur in India through Malaysia to New Zealand coastal waters upto 20 metre depth, and in estuaries. In India the species are found in northern regions of both coasts, and form one of the most important fisheries in Bombay and Gangetic Delta. It grow to a maximum size of 80 mm.

Macrobrachium equidens is distributed from Africa to South-west New Guinea. In India it is only found in small numbers in fresh and brackish waters of Kerala. They grow to a maximum length of 100 mm. Their rostrum is curved upward and lower margin with five to seven teeth. Their fingers are covered with stiff or velvetty hairs on the entire surface or in the proximal part.

Rostrum of *Macrobrachium idella* is with nine to eleven teeth dorsally, three of which are generally placed behind the orbit. The carpus of second leg in adult male is larger than chela. The species though attain a maximum length of 110 mm form a small fishery in East Africa and Madagascar through Indian coasts to Java, Sumatra and Malayan Archipelago. In India they are present in estuaries and rivers in south-west region and more common in the east coast. From sea stray records are available.

Macrobrachium javanicum form a small fishery in India through Mayanmar to Malayan Archipelago and Borneo. In India they are available in freshwaters and estuaries in deltaic Bengal. They attain a maximum size of 100 mm. The fingers of second legs of adult male are with one or two fairly large teeth. Smaller teeth are present between the first tooth and the base of the fingers. The anterior tooth of the dactylus is placed in or slightly before the middle of the finger.

Macrobrachium lamarrei form a small fishery in India and Pakistan in fresh and brackish waters. They grow to a maximum length of 130 mm and mostly found in north-east coast of India, Chilka lake and West Bengal. The basal crest in the species is not much elevated and provided with five to nine teeth. Palm of the second leg is not swollen and fingers shorter than palm.

Macrobrachium malcomsonii have distinctly elevated basal crest and provided with five to nine teeth. In younger specimen the second leg has swollen palm and the fingers longer than the palm. Carpus of second leg in adult male is shorter than chela. The species is available in India and Mayanmar. In India they occur in peninsular rivers that drain into Bay of Bengal and in Indus river in west. The species migrates into brackish waters during breeding season and forms a fairly good fishery in north east coast in monsoon months. The males and females of the species attains a maximum size of 230 mm and 200 mm respectively.

Macrobrachium mirabile have conspicuously, about one and one third, longer fifth legs than the fourth leg. Rostrum is short and high with many dorsal teeth. Second leg of adult male is smooth. The species is distributed in India through Mayanmar to Malayan Archipelago and Borneo. In India, estuarine species are mostly found in Gangetic Delta where they form a small fishery. They attain a maximum size of 65 mm.

Carpus of the second pereiopod in adult male of *Macrobrachium rosenbergii* is slightly longer than half as long as chela besides the absence of branchiostegal spine. Hepatic spine is present, dactylus of last three legs are simple. Fingers of that leg (2nd pereiopod) are of the same length as the palm. The species is widely distributed in the Indo-Pacific region, the western limit being the Indus delta area and extending upto Indo-China in the Asian mainland. They are found in fresh and brackish water. They are common in Indian lakes and estuaries along the coast lines. The species used to form a very good fishery in the monsoon and post-monsoon months but the fishery is now diminished due to indiscriminate fishing. The species attain a maximum size of 320 mm.

Macrobrachium rude form a good seasonal fishery in Bengal and Orissa. It is available in East Africa, Madagascar, India and Sri Lanka. In India the species occur on south west coast and east coast and common along deltaic Bengal, Orissa and Andhra Pradesh. Larger chela of second leg of adult male of the species are with tubercles at both sides of the cutting edges. Carpus of second leg in adult male is shorter than chela. All joints of second leg in adult male is pubescent. It attains a maximum size of 130 mm.

Macrobrachium scabriculum inhabits around Indian Ocean, extending from Africa through south-west and eastern coasts of India to Malaya Archipelago. It forms a small fishery in deltaic Bengal, Chilka Lake, south west coast of Kerala and south-east coast of Tamil Nadu. They attain a maximum length of 100 mm. The species is recognized by fingers of second leg of adult male with more than four teeth placed at regular intervals, sometimes restricted to the proximal part. Teeth are generally of equal size, but one of the proximal teeth may some times be larger. Fingers are with a velvetty pubescence in their basal portion. Dorsal teeth of the rostrum beginning in the distal third of the carapace.

Macrobrachium villosimanus, though attain a maximum length of 150 mm, they have a very limited distribution and occur in small numbers in Kolkata and Chittagong in India and Bangladesh respectively. They also occur in Rangoon of Mayanmar. Carpus of second pereiopod in adult male of the species is as long as, slightly longer or slightly shorter than chela. Fingers of that leg is a little less than half as long as the palm.

Heterocarpus gibbosus is characterized by carapace with longitudinal carinae on the lateral surfaces. Its integument is very firm and pereiopods of second pair is very unequal. Abdominal terga, though carinated, never produced posteriorly into overhanging spines. Post-ocular carinae is present. Proper upper margin of the rostrum is armed with two or three teeth. The species is distributed off Tablas Island, Indian seas, Bali Sea and Kei Islands. In India they are present in the Arabian Sea, Bay of Bengal and Andaman Sea. They grow to a maximum size of 140 mm. Available in small numbers from deep water trawling off Kerala coast.

Heterocarpus wood-masoni occuring in East African coast through Indian Seas to Kei Islands. In India they are available in the Andaman Seas and in Arabian Sea off Kerala coast. The species attain a maximum size of 130 mm and is considered as potential commercial importance. In deep-water trawling this species was predominant in the catch. The third abdominal tergum of the species is armed with an acute spine arising from the anterior half. Post-ocular carina is completely wanting.

Pereiopods of *Parapandalus spinipes* are without epipods. Upper margin of rostrum is finely and evenly serrated along its whole length. Carpus of fifth leg is shorter than propodus. Sixth abdominal somite bear minimum thickness, when looked at dorsally two fifth length of this somite. Telson is almost one and half as long as fifth somite. The species occur at Zanzibar through Red Sea, Gulf of Aden and Arabian Sea to Malay Archipelago and New Guinea. In India the species is recorded in the Arabian Sea off the south-west coast off Tuticorin. The species grow to a maximum size of 130 mm and considered as potential commercial importance. Obtained in fairly good quantities in deep water trawling from 150 to 200 fathoms depth off Kerala coast and at 250-400 fathoms off Tuticorin coast.

At least the first two pereiopods of *Plesionika ensis* are with epipods. Posterior lobe of scaphognathite is truncate. Stylocerite pointed anteriorly. Posterior border of third abdominal tergum acutely produced into a sharp tooth that overlaps the next tergum. The species has a world-wide distribution, Pacific, Atlantic and Indian Oceans. The species occurs in Barbados, Martinique and Granada of West Indies, Barra Granada in Brazilian coast, Rio Muni in Gulf of Guinea, Hawaiian islands and Australian Sea. In India, from Arabian Sea the species is available in deep water areas off south west coast. It attains a maximum length of 125 mm. Specimens are obtained from deep sea trawling off Kerala Coast.

Plesionika martia is widely distributed in Pacific and Atlantic regions, Eastern Atlantic and Mediterranean through Indian Seas to Japan, Australia and Hawaiian Islands. In India the species is found in the Arabian Sea and Bay of Bengal. Its posterior border of third abdominal tergum though convex is not acutely produced. Its rostrum is 45 to 67 per cent of the length of the body from orbit to tip of the telson. The species attain a maximum length of 125 mm. The species is considered to have potential commercial importance. Caught in good numbers from deep water trawling off Kerala coast at 150 to 200 fathoms depth.

Aristaemorpha wood-masoni occurs only in small numbers in Bay of Bengal, Andaman Sea, Arabian Sea. The species has also been recorded from south east Australia. In India the species is obtained from deep waters of 180 to 271 fathoms. Attain a maximum length of 120 mm. The rostrum of the species bear many teeth on upper border. Hepatic spine is present. The length of pterygostomian region is more than two and half times its greatest breadth.

Aristeus alcocki can be identified with three-toothed rostrum dorsally. Hepatic spine is absent. Pleurobranchiae on segments ten to thirteen is reduced to mere papillae. Occur in Gulf of Aden, Bay of Bengal, Arabian Sea near Laccadives and Cape comorin. It grows to 150 mm of maximum length. A few numbers were caught from exploratory trawling off south-west coast of India.

Considered to be potentially important in commercial fisheries, *Aristeus semidentatus* appear in Kermadec Islands, Kei Islands and Arabian Sea. In India, the species is recently discovered off Cochin and Alleppy, south-west coast. They grow to a maximum size of 150 mm and obtained in fairly large numbers in deep water trawling at 150 to 200 fathoms off Kerala coast. They can be identified with the presence of pleurobranchiae on ten to thirteen segments on distinct filaments provided with pinnules.

Aristeus virilis, characterized by the presence of pubescent integument occur only in stray catches from Andaman Sea through East Indian archipelago to Japan. In India, the species is available from Andaman Sea at 188 to 405 fathoms. The species attain a maximum length of 150 mm.

Atypopenaeus stenodactylus is found in Indian seas through Malaysia and Hongkong to Japan. In India, the species is available on east and west coasts. The species form a fishery in Mumbai and are caught in bag nets (Dol nets) from 6 to 15 fathoms in large numbers throughout the year. Small numbers are available in east coast. The species attain a maximum length of 50 mm. The carapace of the species is without longitudinal sutures. Ischial spine is present on second pereiopod. Hepatic spine is present. Petasma is not constricted distally. The anterior plate of thelycum is rounded posteriorly.

Atennular flagella of *Hymenopenaeus acqualis* is cylindrical or sub-cylindrical. Its rostrum is straight, inclined upwards at an angle of 20° with 7-8 +2 spines dorsally. The species is available in east coast of Africa along Indian seas to Japan. In India it is available on the south west coast and Andaman sea from beyond 150 fathoms. The species grow to a maximum size of 90 mm. Only stray catches are available.

Metapenaeopsis andamanensis, growing to a maximum size of 135 mm is distributed in Indian seas through Malaysian waters to Kei Islands and Japan. In India, they are available off south west coast and Andaman Seas. The species is considered as potential commercial importance. Fairly good numbers are obtained from deep sea trawling off south west coast of India from 150 to 200 fathoms, off Tuticorin at 250-400 m. depth. Petasma of the species is asymmetrical. Third maxilliped is with basal spine. Posterior extension of thelycal plate with indistinct median sulcus and angular postero lateral corners.

Metapenaeopsis mogiensis characterized by the presence of a pair of tooth-like platelets behind thelycal plate and lacking of posterior tubercles, grow to a maximum size of 90 mm and obtained in stray catches only in India to South China sea and Japan in north east and tropical Queensland and Sri Lanka in south east. In India they are available off Malabar coasts and Andamans. Their petasma is asymmetrical and third maxilliped is with basal spine.

Metapenaeopsis philippine have distinct median sulcus in the posterior extension of thelycal plate and evenly rounded postero–lateral corners. Their petasma is asymmetrical and third maxilliped is with basal spine. They are distributed in Zanzibar through Indian Seas to Philippine islands. In India they are caught from south west coast. The species grow to a maximum size of 130 mm and is considered as potential commercial importance. They are caught in deep sea trawling off Kerala coasts in good numbers.

Metapenaeopsis stridulans are distributed in Indian, Sri Lankan through Malaysian waters to eastern New Guinea. In India they are available in Bombay area and northern area of east coast, at 5 to 30 fathoms depth. They grow to a maximum length of 100 mm and forms a fishery in Bombay, caught in 'dol' nets in fairly large numbers especially in October and November. Dorsal carina of third pleonic somite of the species is sulcate. It's stridulating organ is almost straight. The anterior edge of thelycal plate is entire. Its left petasmal lobe is sharply pointed and triangular.

Metapenaeus affinis, characterized by distinct branchicardiac carina, extending from posterior margin of carapace almost to hepatic spine, has anterior thelycal plate longitudinally grooved, wider posteriorly than anteriorly. Distomedian petasmal projections are crescent shaped. The species is distributed in Indian Seas through Malaysia and parts of Indonesia to HongKong and Japan. In India, it is found along the coasts and form a very important fishery along the coasts, majority are caught from in-shore waters upto 45 to 50 metre depth. Juvenile enter the estuaries and are fished from there. The species attain a maximum size of 180 mm.

Posterior extension of the anterior median thelycal plate of *Metapenaeus alcocki* is bound laterally by an oval plate on each side. Distomedian petasmal projections are overlying lateral projections and distally trilobed. They grow to a maximum length of 97 mm and are caught in very small numbers. This new species are available in Gulf of Kutch only.

Metapenaeus brevicornis form a good fishery in northern regions of south and east coasts. Juveniles enter into the estuaries from where they are caught. The species is distributed in west Pakistan through India, Malaysia, Thailand and Indonesia to about East Borneo. In India the species is available in the northern region of both coasts. Juveniles are found is estuaries. They grow to a maximum length of 125 mm. Posterior part of rostrum is with distinctly elevated crest. Basal spine is simple found on third pereiopod of the male. Apical petasmal filaments are slender and slightly converging. Thelycum is with a large anterior and small lateral plates.

Metapenaeus burkenroadi is distributed north of equator from Japan through HongKong and Malaysia to south India and Sri Lanka. In India the species have been recently been recorded from inshore waters and estuaries of Cochin. Though the species attain a maximum size of 100mm, only a few have been recorded recently. Branchiocardiac carina of the species is feeble or ill-defined. Anterior end not exceeding posterior one-third of carapace. Distal margin of anterior thelycal plate is convex to indistinctly triangular. Petasma is found with laminose and strongly diverging distomedian projections.

The posterior part of rostrum of *Metapenaeus dobsoni* is without distinctly elevated crest. Basal spine on male third pereiopod is long and barbed. Apical petasmal filaments are not really visible. Anterior thelycal plate is tongue-like. The species is distributed in India through Malayasia and Indonesia to Philippine islands. They are found in estuaries also. In India, the species is more common along south west coast upto 40 metre depth. The species attain a maximum length of 125 mm and form one of the most important commercial fisheries, a major fishery along south-west coast. Juveniles of 70 to 75 mm size are fished from estuaries and river mouths.

Metapenaeus ensis, caught on the east coast of India, where the species form a small fishery along with *M. monoceros.* They attain a maximum size of 170mm and distributed in Indian waters, Sri Lanka through Malacca Strait and Indonesia to New Guinea, south east China to Japan, South western Australia, Queensland and New South Wales. In India the species is available on east coast of Waltair and a little south. Distomedian petasmal projections of the species are directed anteriorly. Lateral thelycal plates are with raised lateral ridges, each with a posterior inwardly curved triangular plate.

Metapenaeus kutchensis, recognized by posterior extension of the anterior median thelycal plate not bound laterally by oval plate on either side and distomedian petasmal projections not overlying lateral projects. The species was recorded only from Indian waters in Gulf of Kutch where it contributes to a fishery. The species attain a maximum size of 140 mm in length.

Metapenaeus lysianasa is found in India, Sri Lanka to North Borneo. In India it is mostly available in east coast off river Hoogly, Orissa and Gulf of Mannar, The species is very rare in south west coast. It attains a maximum size of 90 mm and only available in stray catches. Rostrum of the species is wide and short, not reaching to distal end of basal antennular segment. Thelycum is with ovoid anterior and lateral plates of sub equal size. Conjoined pads are usually set askew. Apical filaments of petasma are vestigeal represented by a pair of rounded bossess.

Lateral thelycal plates of *Metapenaeus monoceros* are with salient and parallel ear-shaped lateral ridges. Distomedian petasmal projections are hood–like. The species attain a maximum size of 180 mm and forms a very important fishery of commercial importance. The species is distributed to South Africa through Mediterranean and Indian Seas to Malaysia. Eastern limit of the species is Malacca Strait. In India the species is available along the entire coast line. Adults are found in the deep sea at 50 to 60 metre depth and juveniles in the estuaries.

Metapenaeus stebbingi are available in Eastern Indian Ocean region, western South Africa through Suez, Red sea to West Pakistan and north west coast of India. The species was recorded recently from Gulf of Kutch. It attains a maximum length of 120 mm and found in stray catches only. Branchiocardiac sulcus of the species is almost completely absent. Distomedian petasmal projections, are anteriorly filiform each with a serrated ventral margin.

Parapenaeopsis acclivirostris is distributed in South Africa, Persian Gulf and Indian Seas. In India they are available in east and west coasts, and found in small numbers with other commercial species. The species attain a maximum length of 50 mm. Anterior plate of thelycum is with a more or less straight transverse posterior edge. No accessory ridges are found on anterior edge of posterior plate. Rostrum is inclined upwards at an angle to carapace for whole of its length.

In *Parapenaeopsis maxillipedo* petasma is with a pair of long slender caliper-like disto-lateral projections directed forwards. Thelycum is with median tuft of long setae behind posterior edge of last thoracic sternite. Third pereiopod of female is provided with basal spine. The species is distributed over equatorial region, spread over from west coast of India and Sri Lanka through Malaysia to Philippines and New Guinea. In India it is recorded from Bombay, Kerala and Madras coasts only in small numbers. The species attain a maximum length of 125 mm.

Antennular flagella of *Parapenaeopsis hardwickii* are 0.7 length of carapace or longer. Thelycum is without a median tuft of setae on posterior plate. The species is available in Indian waters through

Malaysia to South China. In India it is found in Bombay area and off Godavary estuary. It attains a maximum length of 120 mm and form a fairly good fishery in Godavary estuary and more in Bombay coast.

Post-rostral carina of *Parapenaeopsis nana* reaching three fourth carapace. Petasma with a pair of disto-lateral projections directed laterally. Cup-like distal projections are absent. The species is found in Indian and Sri Lankan waters. In India, it is recorded only from east coast off Orissa and Chennai coasts where it is obtained in stray catches only. The species grow to a maximum length of 55 mm.

Parapenaeopsis sculptilis attain a maximum size of 165 mm and form a small fishery in northern east and west coasts of India. The species is distributed in India through Malaysia, Indonesia to HongKong and Australia and New Guinea. In India the species occur in northern areas of east and west coasts. Antennular flagella of the species are 0.5-0.6 length of carapace. Its thelycum is with median tuft of setae on posterior plate.

Telson of *Parapenaeopsis stylifera* is with a pair of fixed subapical spines. At least distal half free portion of rostrum is unarmed. The species is available in India and Sri Lanka through Malaysia to Indonesia and Borneo. In India it is available all along the coast lines, more along west coast and north east coast. There is no estuarine phase in its life history. It grows to a maximum length of 140 mm and is considered a very important commercial species along the entire west coast of India with maximum abundance along the Kerala coast.

Anterior plate of thelycum of *Parapenaeopsis tenella* is with V-shaped posterior edge; and two accessory ridges on anterior edge of posterior plate. Rostrum of the species is with proximal one-third rising from carapace. The remainder is more or less horizontal. The species occur in east coast of India and Mayanmar through Malaysia to Northern China, Southern Japan and Northern Australia. In India, the species is recorded from Palk Bay and Gulf of Mannar only in small numbers. It grow to a maximum size of 50 mm.

Parapenaeopsis uncta occurs in India, Sri Lanka and Malayan waters. In India it is recorded from Orissa and south-west coast only as stray catches. It grow to a maximum length of 100 mm. The second pereiopods of the species is without basal spines.

Branchiostegal spine of *Parapenaeus fissurus* is on anterior margin of carapace. Sixth abdominal somite is less than twice the length of fifth. Process "a" of petasma bifurcate and directed laterally. Thelycum is with anterior, intermediate and posterior plates. The species is distributed in East Africa through Indian Seas, Malaysia and Indonesia to Philippines, South China Sea and Japan. In India it is only recorded from the east coast off Ganjam and also in Andamans. It grows to a maximum length of 120 mm and form stray catches only.

Parapenaeus investigatories is distributed in East Africa through India and Indonesia to Japan. In India, the species is available mostly in Gulf of Mannar, off Pulicat Lake and Andamans. They are also caught off Cochin in deep sea trawling. It grows to a maximum size of 80 mm and occur in small numbers. The branchiostegal spine of the species is situated a little behind anterior margin of carapace. Sixth abdominal somite is more than twice the length of the fifth. Rostrum is reaching distal end of first segment of antenular peduncle.

Branchiostegal spine in *Parapenaeus longipes* is absent. Fifth pereiopods of the species exceeds antennal scale by dactylus. The species is distributed in East Africa through India and Indonesia to New Guinea and Japan. In India, they occur in Mangalore and Cochin, off Ganjam, Vizagapatnam and river Hoogly. They grow to a maximum size of 80 mm and caught in small numbers.

Penaeopsis rectacuta occur in Gulf of Aden through India and Malaysia to Philippines, South China Sea and Japan. In India they are found off Chennai, Andamans and off Kerala coast. The species grow to a maximum size of 130 mm and is considered as potential commercial importance. It is obtained in appreciable numbers off Kerala coast in deep sea trawling at 100 to 200 fathoms depth. Carapace of the species is without longitudinal sutures. Branchiostegal spine is present. Telson with three pairs of movable marginal spines in addition to the fixed pair.

Telson of *Penaeus canaliculatus* is unarmed and rostrum with one ventral tooth. The species is found in South Africa, Mauritius and Red Sea through India to East Indies and Fiji Islands. In India they are available in west coast, where they are caught in small numbers. It grows to a maximum size of 150 mm.

Gastro-orbital carina of *Penaeus indicus* occupy the posterior two third distance between hepatic spine and orbital angle. Rostral crest may be elevated but not triangular in profile. The species is distributed in India, Sri Lanka to the west through Gulf of Aden and Madagascar and east coast of Africa and to the east through Malaysia and Indonesia to Philippines, New Guinea and northern Australia. In India the species occur in all coastal waters of east and west. Young migrate to estuaries. The species attain a maximum size of 230 mm and form one of the most important commercial species. In estuaries there is a fishery of young ones of 120-140 mm size.

Penaeus japonicus is widely distributed throughout the greater part of Indo Pacific region, from Africa to Fiji. In India it occur along Tamil Nadu coast and on the west coast especially in Bombay. In attain a maxinim size of 270 mm and form a small fishery in Tamil Nadu area, especially in Pulicat Lake in post monsoon months. In Bombay area, they are found in small numbers. Adrostral sulcus of the species is narrower than post-rostral carina. Anterior plate of thelycum is rounded at the apex.

In *Penaeus latisulcatus,* adrostral sulcus is as wide as post-rostral carina. Anterior plate of thelycum is bifid at the apex. Scattered distribution of the species have been recorded from Red Sea through Malaysia and Malaccas to Korea and Japan. In India only one or two specimens were caught from south-west coast. It attain a maximum size of 160 mm.

Penaeus merguiensis characterized by triangular rostral crest, adrostral carina not reaching as far as epigastric tooth and dactylus of third maxilliped of adult male is half the size of propodus, is distributed from West Pakistan eastwards to New Caledonia, penetrating southwards to Australia. In India the species is available mostly in the middle of east and west coasts. In other areas they are found in small numbers. Juveniles enter into the estuaries. They grow to a maximum length of 240 mm and form small fishery in the middle region of east and west coasts. Juveniles are fished from estuaries.

Penaeus monodon is distributed from South Africa to Southern Japan. In India it is more common in east coast, especially in Bengal and Orissa coast and all along west coast. Juveniles enter estuaries. This is the largest Indian marine prawn which grow to a maximum size of 360 mm. It is an important commercial species and form substantial fisheries in Bengal and Orissa. In west coast the species forms a fishery, though not as dominant as in the east, but large ones are caught in northern areas of west coast. Young ones are caught from estuaries. Hepatic carina of the species is horizontally straight and its fifth pereiopod is without exopodite.

Dactylus of third maxilliped of adult male *Penaeus penicillatus* is much longer than propodus. Adrostral carina of the species reaching just beyond epigastric tooth. Its rostral crest is markedly elevated. It is distributed from Karachi coast in West Pakistan through Malaysian waters to Taiwan. In India the species is found in Bombay and Orissa coasts. It attains a maximum size of 210 mm and small numbers occur in Bombay and Orissa.

Penaeus semisulcatus is widely distributed in Indo-West-Pacific, from Durban Bay through Red Sea, India, Malaysia, Indonesia to northern and north-eastern Australia through New Guinea, Philippine Islands to Southern Japan, mostly in tropical habitats. In India, the species is more common on the east coast. It grows to a maximum length of 250 mm and form a small fishery in east coast. Hepatic carina in the species is inclined at an angle of 20 degree anteroventrally. Fifth pereiopods are provided with small exopodite.

Post-rostral carina of *Sicyonia lancifer* is armed with five teeth. Abdominal pleura of first and second segments are unispinose and third, fourth and fifth are with three spines. The species is available in Japan, Penang, Gulf of Mannar, Sri Lanka and India. In India they occur on the south west coast at 12 to 17 fathom depth. It grows to a maximum length of 80 mm and only a very small numbers are caught from Arabian Sea.

Stray catches of *Solenocera choprai* occur from Arabian Sea from a depth of 56 to 58 fathoms. The maximum size attained by the species is 130 mm. Spine on the cervical groove ventral to the posterior-most spine of the rostral series are absent. 'L'-shaped groove on either branchiostegal region is absent.

Solenocera indica is distributed from Bangladesh along Indian coasts and Sri Lanka to Malayasia, Borneo and Hongkong. In India the species is available along east and west coasts at 40m depth or less. It attains a maximum size of 130 mm and forms a small fishery in Bombay. Telson of the species is simple and devoid of any spine on lateral margin.

In *Solenocera hextii* spine in cervical grove ventral to posterior most spine of the rostral series is present. 'L' -shaped grove on either branchiostegal region is also present. The species inhabit in northern Indian Ocean regions. In India it is available along entire west coast and Bay of Bengal at 65 to 276 fathoms depth. It grows to a maximum size of 140 mm. During deep water explorations at 150 to 200 fathoms depth, the species was caught in varying numbers, but never in large quantities. The large sized specimens is attractive to commerce.

Solenocera pectinata has externo-distal margin of the exopod of the uropod without spine. Post-rostral carina is not extending beyond cervical grove. The species is distributed in Arafura Sea, Flores Sea, Ceram Sea off Owase, Japan, Arabian Sea Tennesseram coast, Mayanmar and South China Sea. In India the species is available off south west coast at 25 to 60 fathoms. They grow to a maximum length of 75 mm and only found in small numbers. Their size is small and commercially is not attractive.

Trachypenaeus curvirostris is distributed in Eastern Africa through India, Srilanka and Malayasian waters to Japan and Australia. They are found at 10 to 30 fathoms in both east and west coasts of India. They do not occur in large numbers to contribute to a fishery, and grow to a maximum size of 95 mm. Third pereiopod is with epipodite. The anterior plate of the thelycum may have a raised anterior margin but laterally the margins are not raised. An excavation present between the anterior plate and the transverse sternal ridge.

In *Trachypenaeus pescadorensis* epidodites are absent on first and second pereiopods. Distolateral projections of petasma are with sharp tips reaching coxae of fourth pereiopods. Anterolaterally with large wing-like flaps on outer curvature. The species is available in south-west and south-east coasts of India, eastern Malaya and northern Australia. They grow to a maximum length of 90 mm and stray catches were only obtained from south-east and south-west coasts of India recently.

Trachypenaeus sedili was recorded from Malaya and Sri Lankan waters. In India it is obtained from Bay of Bengal recently in Visakhapattanam. They grow to a maximum size of 60 mm. The species has the plates of thelycum with raised anterior and lateral margins.

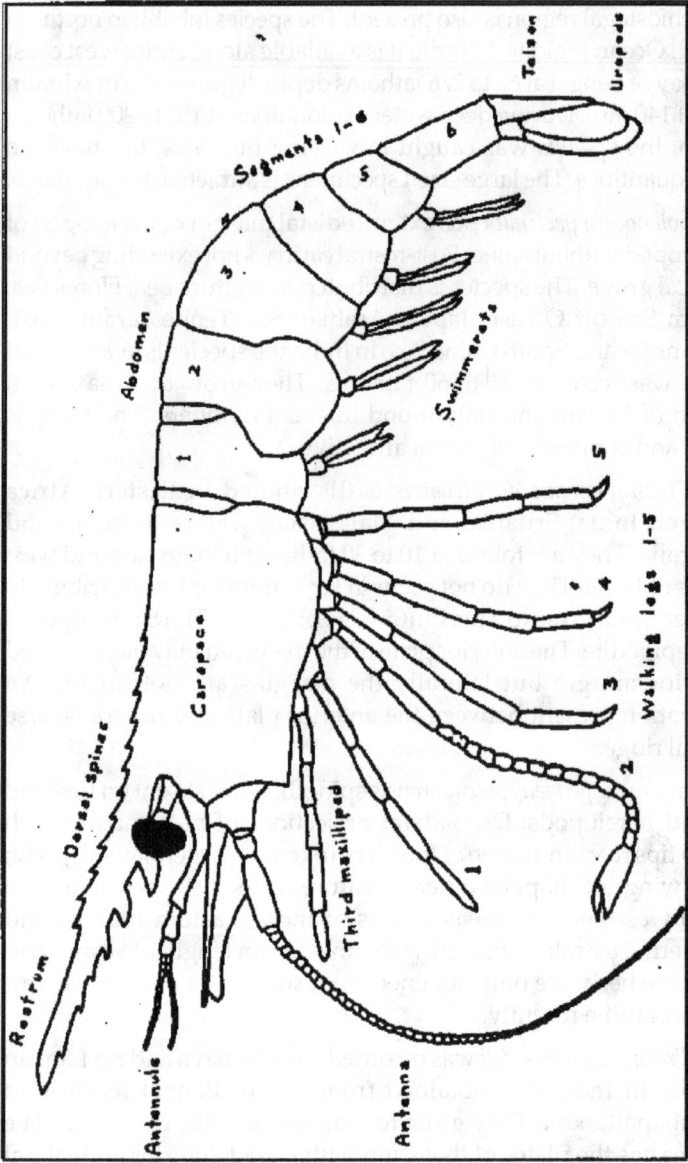

Figure 9: Diagram of Pandalid Shrimp, Showing Body Features

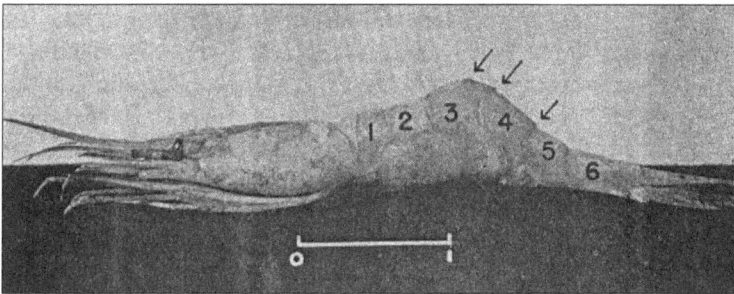

Figure 10: *Pandalus borealis*–Spines 2nd, 3rd and 1 on Fourth Abdominal Segments

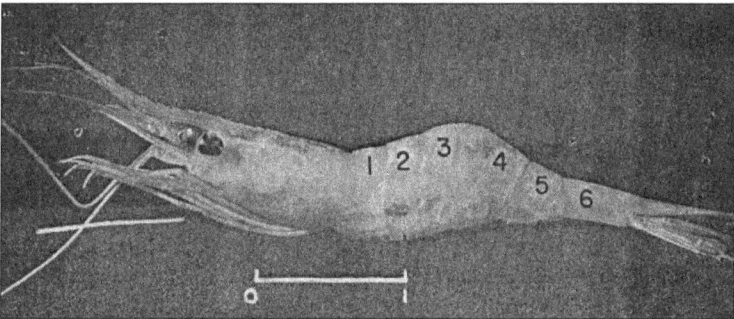

Figure 11: *Pandalus jordani*–Spines Absent in 3rd and 4th Abdominal Segments

The new species, *Acetes cochinensis* was recorded from Cochin waters in India.They grow to a maximum length of 30 mm and occurs in plankton of sea and back waters in May and June. In the species ciliated and non-ciliated portions of external border of exopod of uropod is seperated by a tooth. Distal portion of pars externa are with tubercles.

Acetes erythraeus trochanter (basis) of third pereiopod is without tooth on inner free margin. Petasma with a pair of folded coupling membranes armed with hooks. It is available in Red Sea through Bay of Bengal and Gulf of Siam to Malay Archipelago. In India the species is found in east and south-west coasts. It grows to a maximum size of 30 mm and form a fairly good fishery in Bengal, Orissa and Tamil Nadu. In Trivandrum coast the species enter in large quantities from December to April.

Acetes indicus in distributed in India through Mergui Archipelago and Gulf of Siam to Malaya and East Indies. In India, it is common in northern area in west and in the east throughout the coastal and estuarine regions. The species grows to a maximum length of 40 mm and from good fishery in north west coast and east coast. In Bombay the species contribute 20 per cent of prawn fishery. The trochanter (basis) of the third pereiopod is with tooth on inner free margin. Petasma in without membraneous coupling folds.

Acetes japonicus occur in India, Gulf of Siam, Java, Korea and Japan. In India the species is available in west and south east coasts and lower parts of Bay of Bengal. The maximum size of the species is 26 mm. It occurs in small numbers off Trivandrum coast, large numbers are found in July. Ciliated and non-ciliated portions of external border of exopod of uropod is not seperated by a tooth in the species. Distal portions of pars externa are with tubercles.

Apex of telson is truncated in *Acetes serrulatus* with a tooth in each corner. Segment preceding the one bearing the clasping spines is with angular process pointing backwards. The species is available in Indo-China Sea, Borneo, Signapore. In India, only the variety, *A. serrulatus johni* was recorded from south west coast. It grows to a maximum size of 23 mm and form a fairly good fishery in coastal waters of Travancore (South Kerala) from December to April.

Acetes sibogae is distributed in India, Bay of Bima and Java Sea, Malaya and New South Wales. In India stray catches are available off Quilon in south west coast. It grows to a maximum size of 35 mm. Apex of telson is triangular in the species. External antennula flagellum in male is with single clasping spine.

Pandalopsis dispar, commonly known as "Side-stripe" or "Giant-red" is distributed on the Pacific coast from Bering sea to Washington coast. Their antennules are twice as long as carapace and red and white stripes are present on each side of abdomen. A large specimen may measure 200 mm from the tip of the rostrum to the tip of the telson. The extremely long antennules and striped abdomen easily distinguish this species from other commercial shrimps. The species is found on muddy bottoms and is generally trawled at depths ranging from 100 to 120 metres.

Pandalus borealis, commonly known as "Pink" shrimp are characterized with median spine on dorsal (upper) surface of third

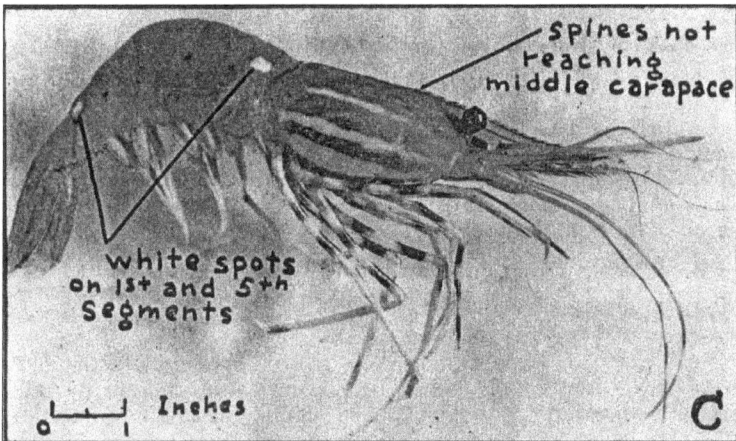

Figure 12: Prawn (*Pandalus platyceros*)

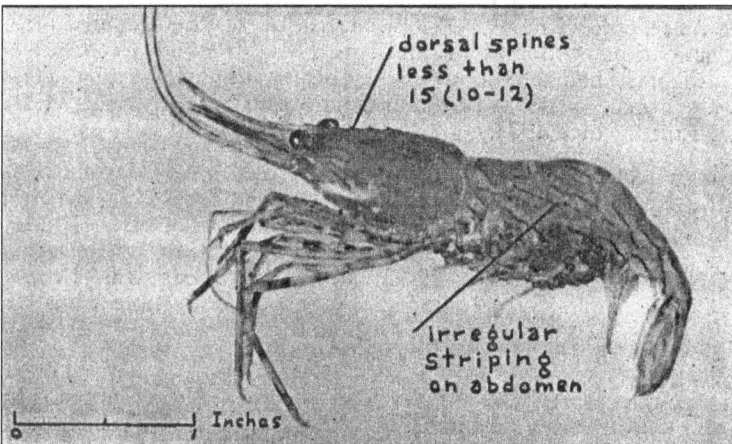

Figure 13: Coon-stripe (*Pandalus danae*)

abdominal segment, pointing to rear and no horizontal striping on walking legs, besides absence of red and white stripes on abdomen and antennules less than twice as long as carapace. Though the species, in general, is 75-100 mm in length, but larger individuals may reach 150 mm and landed in bulk quantities. The "pink" shrimp is the only commercial species lacking the white horizontal striping on the walking legs. The sharp spine or lobe pointing backward on the third abdominal segment is the most distinctive feature of this

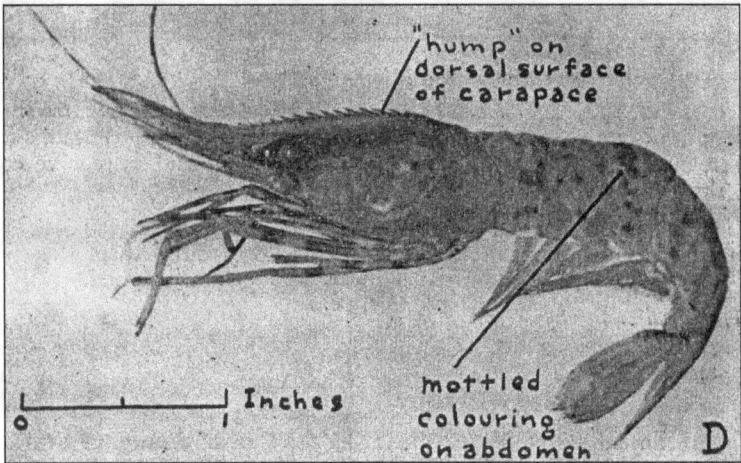

Figure 14: Humpback (*Pandalus bypsinotus*)

species. The distribution of this shrimp is circumpolar, ranging from the Columbia river on the Pacific coast to Massachusetts bay on the Atlantic coast of America. They form the basis of the fishery of outer Burrard inlet where it is fished on the muddy bottom at depths ranging from 20 to 120 metres.

Pandalus platyceros ("Prawn" or "Spot") is the largest of the commercial shrimps. Large individuals may reach almost 225 mm from the tip of the telson to the tip of the rostrum. The 14 to 17 dorsal spines extend from the middle of the rostrum, but do not reach the middle of the carapace. The colour of the body is usually reddish brown with distinctive white spots on the first and fifth abdominals segments and with horizontal white bars on the carapace. The "prawn" as an adult is generally found on rocky bottoms, and so is not usually caught in quantity by trawling. It has been successfully trapped in some large mainland inlets at depths from 80 to 140 metres. The species is distributed on the Pacific coast from Alaska to San Diego, California.

The "Hump-back" shrimp (*Pandalus bypsinotus*) is given its common name because of the arched shape of the dorsal surface of the carapace. The dorsal spines, of which there are 17 to 21 extend to the middle of carapace. The species may reach a length of 150 mm. The body is coloured with a mottling of reddish brown which is conspicuous on the abdomen. The species is distributed from the Bering sea to the strait of Juan deFuca.

Pandalus danae, the "Coon-stripe" shrimp derives its common name from the irregular striping on the abdomen. The colour of the body is made up of varying sheds of brown and red. The dorsal spines which extend behind the middle of the carapace number 10 to 12. The species is found on a sandy bottom, usually where a rapid tidal current exists. They are distributed from Sitka, Alaska to San Francisco, California. A large specimen may attain a length of 125 mm.

Chapter 2
Habitats Around Indian Seas

Mixing processes in the oceans have great significance to the biological productivity because of the high fertility they cause in the euphotic zone. The biological cycle is clearly influenced by the process of vertical mixing processes (upwelling) and wind pattern (monsoon gyre) and driving force of the currents creating convergence zone.

Wind Type

The northern Indian Ocean falls within the monsoon gyre, where predominant wind direction changes gradually from northeasterly to northwesterly and later to westerly during March to May. In June, the direction changes to southwesterly and westerly with increasing velocities and by July, the velocity reaches Beaufort 5 to 7. During June to August, the predominant wind direction is from southwesterly to westerly. By the end of August, the wind velocity decreases and by October-November the wind starts blowing from northwesterly to east northwesterly with comparatively low velocities. Both sea and land breeze are common in this area except during south west monsoon (along the Indian west coast) and the north-east monsoon season (along the Indian east coast)

Currents

The circulation pattern in the Arabian Sea and Bay of Bengal comprise of the monsoon current, the equatorial current and the equatorial counter current. The monsoon current is westerly during the northeast monsoon period and easterly during the southwest monsoon period. The north equatorial current is westerly and the equatorial flow easterly. The coastal circulation becomes clockwise during the southwest monsoon (May to October) and counterclockwise during the north-east monsoon season (October to December). Eddies occur within a current system or between two current systems under the above current pattern. During the southwest monsoon season, the northeasterly current do not exist and the equatorial current and monsoon current dominate the area. In general, the speed of the current is greater in Bay of Bengal and the Arabian Sea compared to the equatorial regions of the northern Indian Ocean, except off the central part of the east coast of India, particularly in the months of March and April. The strongest currents are generally found off the Somali coast.

During the southwest monsoon, there is southerly flow along the Indian west coast spreading over the entire shelf region. During the change from the southwest monsoon to the winter, the northerly current is established off the shelf. Adjacent to and on the seaward side of northerly flow is present a southerly current limited to the southerly region. From the winter to summer (February to April), the northerly current disappear and the circulation breaks up into eddies. The southerly current persists in summer though it is limited to a narrow belt. During the southwest monsoon period, this narrow southerly current spreads over the entire shelf. The winter current appears to be stronger than the southwest monsoon current along the west coast of India.

During the south-west monsoon period (May to October), the surface current flows southwards along the west coast of India, thereby shifting the isolines for the different oceanographic parameters (temperature, salinity, dissolved oxygen and density) near the coast. The denser subsurface waters from the intermediate, subsurface levels are slowly induced upwards (upwelling) along the continental shelf to occupy the left side of the southerly current (near the coast). This results in the comparatively denser, colder and low oxygenated subsurface waters reaching the surface levels near the coast. During this season, oxygen deficient waters cover the whole continental shelf area over the bottom.

The southerly current transports the comparatively high-saline Arabian Sea waters southwards along the west coast, though during the rainy season the addition of freshwaters from rainfall and river runoff causes a significant lowering of the surface salinity near the coast.

During the northeast monsoon period (November to March), the surface current reverses its direction and turns northerly. The northerly current advects low salinity equatorial waters northwards at the surface levels causing a convergence located between latitudes 10° N and 12° N where the high-saline Arabian Sea water sinks below the less-saline equatorial waters. The effects of winter cooling at the surface levels along the convergence zone lead to the process of sinking. During the winter season (January-February) the surface mixed layer covers most of the shelf. Far the coastal areas south of Bombay on the northwest coast of India, the resultant speed of the current is reported to be more than 20 km/day during the southwest monsoon period. From November to January a northward flowing current is observed. This drift appears to be shallow and seems to have little influence on the waters below the thermocline.

The seasonal reversal of wind causes a corresponding change in the flow of the surface waters in Bay of Bengal. During the southwest monsoon season, there is a clockwise circulation noticed in the central Bay. The northerly flow is close to the central Indian continental shelf where the speed reaches 3 to 5 knots. During the northeast monsoon, the surface circulation becomes anticlockwise with lesser speeds.

The coastal circulation of the west coast of India during south west monsoon is dynamically similar to the wind driven eastern boundary currents found elsewhere in the oceans. The currents along the west coast of India have a distinct annual cycle. The current turns equatorward in February, gets stronger with time and is most energetic during July and August. Thereafter it decreases in strength, disappears and reverses its direction by October-November and flows away from the equator during November to January. In June there is a shallow (75m to 100m deep) equatorward surface current below which there are downwelling indications of a pole ward under current hugging the continental slope carrying low saline waters in the south west Bay of Bengal. Both the currents weaken towards the north upto 15° latitude and cease to be noticeable at about 20° latitude.

The width of the surface current is about 150 km and the bottom 50 km. The winds during this period vary between west north west near the southern end of the Indian coast to west south west near the northern end. Early remote-forced upwelling in this area result from Kelvin waves originating in the Bay of Bengal.

During the north-east monsoon, there is a poleward surface coastal current along the west coast of India, along a stretch of 400 km wide near the southern end and approximately 200m deep, carrying low saline equatorial surface water. This current is driven by a longshore pressure gradient which overwhelms the influence of the winds during the season. Part of the driving force for these currents also comes from the Kelvin waves triggered by the collapse of the winds along the east coast of India as the southwest monsoon withdraws and the north east monsoon sets in.

Convergence Zone

The existence of convergence zone in the southeast and north east Arabian Sea is evident from the horizontal salinity gradients observed during the period from January to March. The salinity gradients indicated a zone of convergence between Calicut section in the south and the Kasaragod section in the north, a distance of 150 km. The sea surface salinity increased from 33 to 35 ppt from the Karwar section in the south to the Ratnagiri section in the north, suggesting the existence of a convergence zone over a distance of 220 km.

The variations in the sea surface salinity suggest that the convergence zone exhibits seasonal variations, spreading northwards with the intensity of the northward flow, which carries equatorial waters towards the northern latitudes. During the south west monsoon season, the salinity distribution of at the surface level is not indicative of the convergence zone, mainly due to the effect of rainfall and river run off.

Upwelling

Upwelling along the coast occur mostly along the western coasts of the continents as seen on the coasts of Peru, California, northwest and southwest Africa west coast of India. Exception to this pattern are found along the northeastern. African coast, Somali coast, east coast of Arabia, east coast of India. Localised upwellings are found around the areas of divergences. Whereever upwelling is caused by the monsoon wind system and accompanying currents, the phenomenon is more seasonal. The economic benefits of upwelling are mainly due to the large concentrations of commercially important fishes in these areas.

Upwelling occurs in varying intensities along the east and west coasts of India along with the southwest monsoon. The onset of south west monsoon generates the Somali current, resulting a clockwise circulation in the Arabian Sea, which in turn develops into a relatively strong southerly current along the west coast of India. The comperatively cold, low-oxygenated and denser water from the subsurface is slowly brought upwards along the continental shelf, very near the coast. The depth at which the upward sloping motion of the subsurface water mass begins is dependent on the velocity, direction and duration of the prevailing wind system in a given area, the bottom topography, the prevailing current system at the surface and the vertical stability of the water column. Along the west coast, upwelling sets earlier in the south and progressively shifts to the north. The process commences in the deeper waters as early as February and the effect reaches the coastal surface water by May to July, and continues through August to early September in the south and October in the north. The strong southerly flow with the coast on the left side induces the upward motion of the sub surface water near the coast. The winds blowing with the northerly component parallel to the coast till April, help the process to intensify. The thermocline-climbs up during the upwelling season and reaches the surface in June or July. There is no system of wind generated upwelling during the south-west monsoon period along the west coast of India, where the dense bottom waters approach the surface because of the immediate interplay of the current with the tilting of the sea surface and thermocline. Upwelling occurring around the Lakshadeep Islands during November or December is attributed to the divergence of current systems in the vicinity of the islands. Off Cochin, upwelling starts by mid August, establishes by late September and ends by mid October.

Along the east coast of India, prior to the commencement of south west monsoon, in April, the southerly winds blow parallel to the coast. The coastal currents, flowing northerly throughout the coast are favourable to the development of upwelling along the east coast. In the northern hemisphere, a wind blowing parallel to the sea coast with the coast line to its left, or an off shore wind, will favour the process of upwelling. The prevailing current system and not the wind is regarded as the main cause generating and maintaining upwelling. Even if a uniform current velocity is considered all along the coast, the rise of dense, deep water will be stronger in the north further away from the equator.

The time series sections for selected places in west coast from Cape Comorin to Ratnagiri indicate that the process of upwelling commenced earlier in January or February in deeper waters (170-110m). The intensity of upwelling was higher towards the south especially south of Karwar. Vertical time series sections of both temperature and oxygen were used as indicators of upwelling process, since both these parameters influence distribution of fish.

The peak in the zooplankton biomass normally follows the peak periods of upwelling along the south west coast of India. The time lag is dependent on the incoming radition and the inorganic nutrients in the upwelled waters to promote phytoplankton production. The time lag increases when the monsoon rains continue with a cloudy sky, there by reducing the incoming solar radiation, while the nutrient content would depend on the depth from which the upwelled water originates and the time taken by the watermass to reach the surface. This interval varies from year to year and also from place to place. The zooplankton biomass in the Cochin is at its maximum during July–August and minimum during November and March. It is comparatively high in the areas of intense upwelling (between Kasaragod and Karwar). The biomass in this section during July–August, 1978 was double than that during July–August, 1977. The sea bottom of the narrow continental shelf between Quilon and Cape Comorin is mostly rocky and the average plankton biomass does not indicate a proportionate increase with high intensity of upwelling. During upwelling when the oxygen minimum layer (0.5 ml/l) emerges from 100–150 m depth to the surface, some fish populations including crustaceans move into the shallow waters, while the others move off shore, away from the centre of strong upwelling. At times, the rate of upwelling suddenly intensifies, slowing down the replenishment of oxygen through surface aeration and wind mixing.

Along the west coast of India, there is a rich phytoplankton bloom during the south west monsoon off the Trivandrum coast from January onwards and reaches the peak in May but further north, off Calicut and northwards, the peak is attained in July–August, indicating the commencement of upwelling in the subsurface layer much earlier, even when the current is northerly along the west coast. From September onwards the phytoplankton bloom disappears, indicating the cessation of upwelling from thereon.

The average zooplankton biomass in the midshelf for the southeastern Arabian Sea and the abundance of zooplankton along the southwest coast during the monsoon period indicate a recurring pattern in the zooplankton distribution and abundance in shelf waters. The period from July to September is found to be the peak season for plankton production, with a fairly uniform concentration of plankton beyond the nearshore waters all along the coast. Thereafter until December, a shoreward shift in the concentration of plankton is evident, especially in the south. At the same time, the continuous distribution breaks up, becomes patchy and the overall abundance is greatly reduced to the lowest level in January and February. The biological cycle is found to be clearly influenced by the process of upwelling especially in the central and southern region.

The detailed study on the distribution pattern of size and abundance of commercial penaeid prawns of the south west coast of India revealed, that with the exception of *Metapenaeus dobsoni* all the other species were seen to enter the fishing grounds from deeper zones at the beginning of the season by September. They begin to move out of the fishery grounds into deeper zones from February–March onwards. *M. dobsoni* exhibited this movement in opposite direction.

Habitats and Grazing Grounds

The prawn fishery, like any others, is influenced by the environmental conditions, mainly, the hydrological conditions of the habitat. The limit to the amount of life that a given habitat can support (carrying capacity) is determined by various physico chemical processes, in turn are determined by the terrestrial, atmospheric and oceanographic processes in action. The total biomass of all living organisms present in a given area at any one time is referred to as the standing crop. The most important physical and chemical factors determining the quality and quantity of both the standing crop and the carrying capacity of the marine ecosystem include temperature, hydrostatic pressure, salinity, density, dissolved oxygen, carbondioxide, pH and the nutrients, such as, phosphates, nitrates and silicates. Vertical and horizontal distribution of these qualities in the oceans together with the ocean circulation patterns as influenced by atmospheric and geographical factors, play a major role in deciding the productivity of the marine ecosystems.

Frequent changes in the environmental parameters are quite common in shallow water bodies, like estuaries, and have a profound influence on the species that resides in the estuaries. Water currents, tidal effect, water temperature, salinity and other chemical properties, turbidity, vegetation and the nature of substratum influence the occurrence and abundance of different species in space and time.

The environment in which the prawns thrive has profound influence on their successful breeding, growth and survival. Most of the penaeid prawns begin their life in the open sea and migrate to shallow coastal waters, estuaries and back waters at the post larval stages. They remain in these environments for sometime growing rapidly and return to the sea on reaching adulthood or near adulthood. This ontogenic migrations make them capable of withstanding changes in the environmental conditions. Changes in the environmental conditions are comparatively less in the sea. On the other hand back waters and estuaries show large scale fluctuations in the major hydrographic conditions.

Penaeus indicus and *Metapenaeus dobsoni* migrate to near shore waters during monsoon in Kerala. This is due to the environmental changes that take place during monsoon. *Parapenaeopsis stylifera*, which occupies relatively deeper waters during July–August is caught by trawling in Kerala coast. Probability exists for increasing prawn catch by extending commercial shrimp trawling in the offshore waters during monsoon season.

Correlating the prawn catch with hydrographic conditions of the waters off the west coast of India revealed that comparatively higher catches were obtained when colder denser waters prevailed on the continental shelf, and such condition exist across the continental shelf during the monsoon period. The low temperature during this season is the after effect of the vertical circulation as evidenced by the upward shift of the isopycnals and the thermocline. The formation of mud banks along parts of Kerala coast is the cause of flourishing prawn fishery and showed maximum catches during the south west monsoon period. Thus it may be concluded that the prawn migrate in huge numbers when the colder and denser waters prevailed along the coast.

The isothermal layer and the temperature gradient appear to play a major role in the distribution of prawn fishery. The fluctuations in prawn fishery are in phase with the vertical movements of the depth of thermocline. The detailed study of the topography of the depth of thermocline provides an indication for the possible areas of good prawn catch, as the depth of thermocline varies with the vertical circulation of the waters which in turn influence the nutrient distribution.

Parapenaeopsis stylifera remain mostly within 20 m depth contour during non-monsoon period (September to May). With the commencement of south–west monsoon and the consequent changes in the environmental conditions, *P. stylifera* leave the inshore areas in large numbers and occupy deeper zones. They remain mostly in the 20-40 m depth zone during June and in 40-60 m depth zone during July to September.

It has been observed that in general, the peak period of prawn landings occur during and immediately after heavy rains. *Parapenaeopsis stylifera* dominated the fishery during heavy rainfall along the west coast of India. With the strength of tide flow and other factors there is fluctuations in the prawn catch in the stake nets. The highest catch was usually recorded on the new or full moon day or a day or two later. This is due to stronger tidal currents and good water flow through the nets.

Both juveniles and adult prawns prefer soft muddy bottom enriched with organic detritus. While *Penaeus duorarum, P. japonicus* and *P. indicus* like sandy, bottom conditions, most of the other prawns prefer muddy conditions.

Dissolved oxygen concentrations of the water has great influence on the metabolism and growth of the prawns. The rate of oxygen consumption of *Metapenaeus monoceros* changes with the change of water salinity, the consumption increases with increase in high saline waters, the osmotic adaptations found in the prawns. The oxygen requirement of *Penaeus indicus* is directly related to its body weight, the larger prawns need more dissolved oxygen to grow and survive. The lethal limit of dissolved oxygen for the species varies from 1.49 ml/l to 3.8 ml/l among early juveniles and sub-adults respectively.

The metabolism, growth, osmotic behaviour and reproduction of penaeid prawns are influenced by salinity of the water. The adult prawns are relatively less tolerent to low salinities and most of them are known to attain sexual maturity in high saline sea water. The spawning also takes place only in the sea. *Penaeus japonicus* is known to spawn in salinities of 28 to 36 ppt at a water temperature of 22°C. *Metapenaeus dobsoni* was found to spawn in waters above 28 ppt salinity. Penaeid post-larval forms can withstand low salinities, as such they occur in estuaries and backwaters.

Penaeid prawns can tolerate wide variations in water temperature. Intensive spawning activities of these prawns are associated with higher temperature of water. At 28° C to 30° C water temperature larval abundance were peak in waters off Cochin.

Penaeid prawns are comfortable and can carry on their life processes in waters having alkaline pH from 7.5 to 8.5.

Lack of calcium in the environment is likely to affect the frequency of moulting in prawns and play an adverse role in the growth process. The osmoregulatory mechanisms may also break down due to the defective permeability of the membranes.

The movements of prawn in the marine environment is either as a result of biological instinct or due to the environmental consequences. *Metapenaeus monoceros* were found to travel 4 km south to Cochin in 4 days. *M. dobsoni* was found to travel 5.5 km in 4 days and the others 25 km in 8 days to the north of Cochin. *Penaeus indicus* released in May, 1982 at Cochin, south west coast of India were recovered from Tinneveli coast, south of Tuticorin after having travelled a distance of 380 km in 68 days.

A north ward migration of stained *Metapenaeus affinis* in the sea off south Maharashtra coast was observed during February to March and out of 2294 prawns released, 58 were recaptured in 3 days. They travelled a maximum distance of 34 km in six days, an average of 5.5 km per day.

   The possible large-scale movement of *Penaeus indicus* shoals from the inshore waters of Kerala coast to the coast of Kanyakumari district in Tamil Nadu during the monsoon season was observed from the commercial landings. The movement of marine prawns is highly restricted within the trawling grounds from where they are caught. Except a few, the recovery of all marked prawns were made within 20 km from the areas of release. A significant observation of the mark-recovery studies was the recapture of 13 tagged *P. indicus*, migrated long distances ranging from 184 to 628 km southwards along the south-west coast of India and thence skirting round Cape Comorin to the south-east coast upto Manapad in the Tinneveli district of Tamilnadu. The longest distance covered by the prawns was 380 km (Manapad) and the shortest distance 220 km (Thengapatnam), at speeds varying 2.93 km to 5.58 km per day. Southerly current all along the west coast carried the prawns to as far the south-east coast as ascertained by the drift bottle studies simultaneously done during mark-recovery studies.

Chapter 3

Penaeid Prawns Around World Oceans

Marine prawn production in Asia totalled 2.37 million tonnes in 2008. In 2006 global production of prawn was 3.07 million tonnes and Asian production of 2.74 million tonnes.

China's production of the penaeid shrimp totalled 1.21 million tonnes in 2007. A lower production in China was predicted in 2008 because of bad weather. In 2007, *Penaeus vannamei* accounted for 82 per cent of the total production.

The estimate given by Thailand for 2008 was 0.49 million tonnes of *P. vannamei* and 5000 tonnes of *Penaeus monodon*. *P. vannamei* production is expected to decrease to 0.48 million tonnes in 2009. The production in 2007 was recorded as 4,41,451 tonnes *P. vannamei* with less than 1 per cent of *P. monodon* (3.301 tonnes). In 2009 *P. monodon* production is expected to decline further to 1500 tonnes.

Prawn production in Indonesia was 0.3522 million tonnes in 2007. The estimated production was 0.35 million tonnes in 2008.

In Vietnam 0.32 million tonnes of prawn was produced in 2008, which includes 0.10 million tonnes of *P. vannamei*, accounting for 30 per cent of total production. In 2009, the production is expected to increase by 40 per cent at the expense of areas used for *P. monodon* culture.

Table 1: Important Penaeid Prawns Around World Oceans and Commercially Important Penaeid Species

Species Group	Origin	Common English Name	Scientific Name
Coldwater	North Atlantic and North Pacific	Northern shrimp	*Pandalus borealis*
	North east Atlantic	Common shrimp	*Crangon crangon*
Tropical	Indo-Pacific	Green tiger prawn	*Penaeus semisulcatus*
		Banana prawn	*Penaeus merguiensis*
		Indian white prawn	*Penaeus indicus*
		Giant tiger prawn	*Penaeus monodon*
		Kuruma prawn	*Penaeus japonicus*
		Fleshy prawn	*Penaeus orientalis*
		Western king prawn	*Penaeus latisulcatus*
		Brown tiger prawn	*Penaeus esculentus*
	Western Indian Ocean	Indian white prawn	*Penaeus indicus*
		Giant tiger prawn	*Penaeus monodon*
		Green tiger prawn	*Penaeus semisulcatus*
	Eastern Atlantic	Southern pink shrimp	*Penaeus notialis*
	Western Atlantic	Northern white shrimp	*Penaeus setiferus*
		Northern pink shrimp	*Penaeus duorarum*
		Southern pink shrimp	*Penaeus notialis*
		Northern brown shrimp	*Penaeus aztecus*
		Southern brown shrimp	*Penaeus subtilis*
		Southern white shrimp	*Penaeus schmitti*
		Red spotted shrimp	*Penaeus brasiliensis*
	Eastern Pacific	Yellow leg shrimp	*Penaeus californiensis*
		White leg shrimp	*Penaeus vannamei*
		Blue shrimp	*Penaeus stylirostris*
		Crystal shrimp	*Penaeus brevirostris*
		Western white shrimp	*Penaeus occidentalis*
Freshwater	Indo-Pacific	Giant river prawn	*Macrobrachium rosenbergii*

Figure 15: *Penaeus monodon*

Figure 16: *Penaeus indicus*

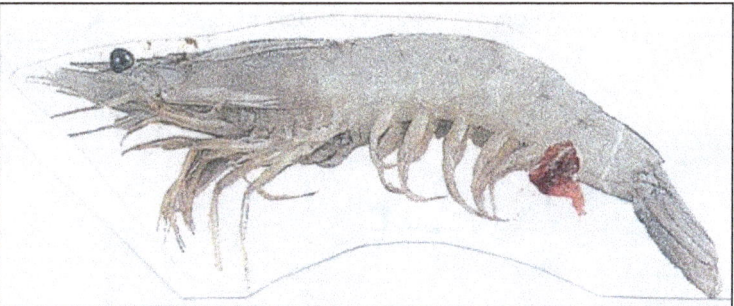

Figure 17: *Metapenaeus monoceros*

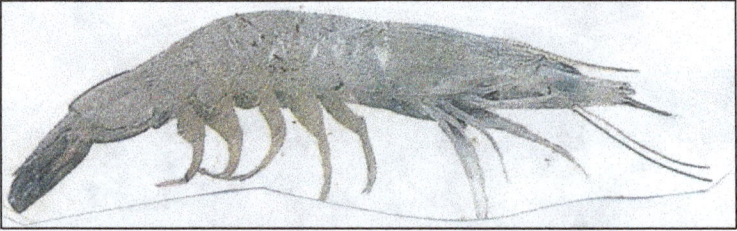

Figure 18: *Metapenaeus dobsoni*

Figure 19: *Penaeus semisulcatus*

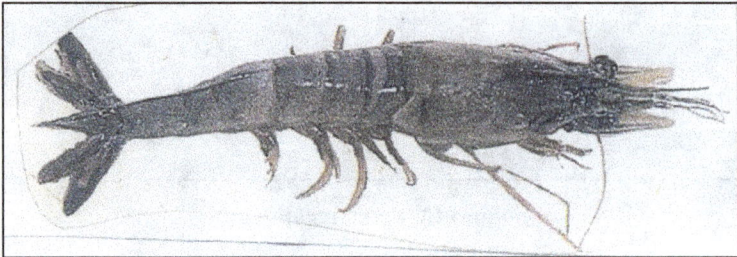

Figure 20: *Parapenaeopsis stylifera*

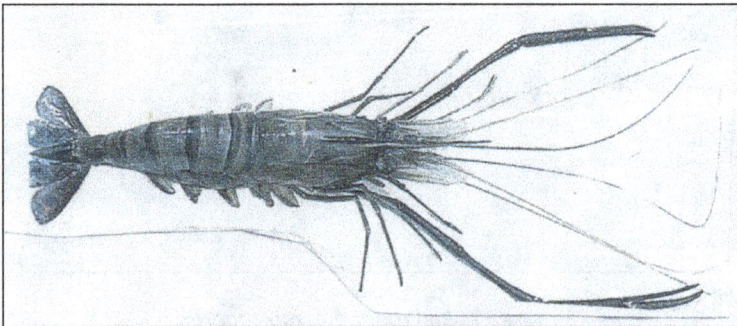

Figure 21: *Macrobrachium rosenbergii*

Prawn production in Philippines during 2007 was 42665 tonnes, mainly of *P. monodon*. Some 4000 tonnes of *P. vannamei* were harvested in 2007. The estimated production in 2008 was 44000 tonnes with less of *P. monodon*.

Prawn industry in Malaysia gave production estimates of 70000 tonnes in 2007, which increased to 80,000 tonnes in 2008. The production could comprise 90 per cent of *P. vannamei*, despite the supply of high health post larvae of *P. monodon*.

Production of prawn was around 60,000 tonnes in 2007 at Bangladesh. Bangladesh was only interested in farming *P. monodon*.

Production of *P. monodon* was 144347 tonnes in India during 2006-2007, which declined by 38 per cent to 106,160 tonnes for the 2007-2008 period. The production in 2008-2009 is expected to be stagnant at 1,00,000 tonnes due to recurring disease out breaks, low farm gate prices, increasing costs of production and rising quality requirements. In 2009 the industry expects 6000 tonnes of *P. vannamei* production in India as the ban on its culture was removed in October, 2008.

Penaeus vannamei prawn is leading in production because of its biological traits. Most wild *P. monodon* stocks are infected. At the same time there was availability of selected SPF and SPR stock of *P. vannamei* from USA, now available in Asia. The survival, production and cost efficiency was higher in *P. vannamei* than in *P. monodon* (20-25 per cent cheaper). *P. vannamei* has better salinity / temperature tolerance. It also had better disease resistance besides having lower environmental impact.

P. vannamei post larvae is cheaper (1.50-4.00 dollar per 1000 PL depending on countries) than that for *P. monodon* PL (2.50–15.00 dollar per 1000 SPF post larvae). The cost of SPF *P. vannamei* brood stock is 30 dollars each, while that of *P. monodon* is 150 to 300 dollars each. *P. vannamei* growth pattern is steady 20 g in 100 days from SPF post larvae, where as *P. monodon* shrimp attain 25 g in 125 days. Nocturnal and active feeding patterns in the *P. vannamei* allows for a much higher stocking density (80 to 175 PL/m^2) in contrast to 5-25 PL/m^2 in case of *P. monodon*.

All these advantages make it difficult for any farmer to resist culturing of *P. vannamei*. In Vietnam, the short growth period of *P. vannamei* has allowed for better utilization of ponds with

P. vannamei culture followed by the seabass. In Asia, producers are also culturing with success large *P. vannamei*.

The choice of *P. vannamei* or *P. monodon* also depends on markets. With high volumes of from *P. vannamei*, it is easy to fill up containers. In contrast, collecting the same volume of *P. monodon* shrimp would require more than three times the number of farms.

In India, there have been declines in farm gate price for the *P. monodon* since 1999. But recently prices dropped as low as Rs. 230-240 (US dollar 4.72/kg) per kg for large shrimp, despite the exchange rate of Rs. 50/- to the dollar. Offer prices in the US markets have been declining at an average of 15 to 20 per cent. The prognosis is to be prepared for lower costs and it is the least cost with *P. vannamei*. The current pricing is favouring the *P. vannamei*. In Vietnam margins are higher with *P. vannamei* culture. Culturing of *P. monodon* is unprofitable as production costs is high and selling prices have dropped drastically (50 per cent).

In India the industry is of the opinion, that the producers will have to do more of value adding, certification for organic aquaculture and species diversification with the introduction of *P. vannamei*.

Table 2: Production of Prawns in 2006 and Estimates for Production in 2008 and 2009 in Tonnes

Country	2006	2008	2009	% of *P. vannamei*
China	1240385	1,000000	1000000	82 per cent
Thailand	500800	490000	392000	99 per cent
Indonesia	339803	276000	300000	89 per cent
Vietnam	349000	320000	350000	30 per cent
India	144347	100000	100000	0 per cent
Malaysia	34973	80000	80000	90 per cent
Philippines	40654	44000	60000	10 per cent
Bangladesh	64700	na	na	
Others (Taiwan, Brunei etc)	23713			
Total	2738375	2370000	2342000	
P. vannamei	1719237	1708080		
Latin America	334260	454000	na	
Global total	3072635			

Prawns and shrimps form the most economically significant group in the marine fishery resources of Indian Ocean.

Bangladesh

Bangladesh bordering Indian Ocean lies within Lat. 20.4° N–22.0° N and Long. 89.0° E–92.0° E. The continental shelf of Bangladesh covers an area of 69900 sq km of which 37,000 sq km is no deeper than 50 m. The bottom materials of the waters up to 40 m depth are mostly alluvial silt and mud. Sand bottom occurs in deeper waters. The estimated standing stock of shrimps in continental shelf was calculated as 9000 tonnes.

In the Bay of Bengal at 30 m depth, 0.7 per cent shrimp were caught during survey and exploratory fishing. From the Meghna river estuary and adjoining Bay of Bengal the following species of prawns have been caught. They are *Palaemon tenuipes, Macrobrachium mirabilis, Palaemon styliferus, Parapenaeopsis sculptilis* and *Palaemon* sp.

Palaemon styliferus inhabits shallow coastal waters and brackish waters of the estuaries. The maximum size of the species recorded is 100 mm. Their breeding period extends from October to July. Hatching of the species occurs in the more saline zone of the estuary or inshore waters.

Adults of *Penaeus indicus* inhabit the sea, whereas post larvae and juveniles are carried into the estuary and inshore waters by the tide and current. Due to its great abundance both in the estuarine and open sea it is considered as one of the most important species of shrimps. It is a bottom feeder and omnivorous. The species is reported to have a deep water spawning habit.

Penaeus monodon is a large 300 mm prawn and one of the most important commercial shrimps of Bangladesh. Post larvae and juveniles are reported to be available the year round from the estuarine area of Bangladesh. Adults and mature specimens are abundantly available in the Bay of Bengal off Cox's Bazaar. The species is omnivorous and a bottom feeder. The breeding period of the species extends from January to April. The spawning ground of the species seems to be the off shore of Cox's Bazaar. Chakria-Sundarban and Khulna–Sundarban estuaries are known as the main seed collection centres in Bangladesh.

Metapenaeus monoceros inhabits sea and brackish water zones of the estuary. Juveniles are found in estuaries and backwaters of reduced salinity; adults occur in the sea. The species are reported to breed throughout the year with two peaks, one in July and August and the other in November and December. The adults are generally available in the sea in slightly deeper waters than the other species of *Metapenaeus*. It is recorded to grow upto 155 mm from Bangladesh waters

India

Distributed along the entire coast of India upto the depth of 50m, juveniles of *Penaeus indicus* occur in estuaries and backwaters. The size of the female at first maturity is 130.2 mm. Fecundity ranges from 68000 to 7,31000 ova in females measuring 140 mm and 200 mm respectively. At Tamil Nadu peak spawning activity was observed from May to September. Feeds on both vegetable and animal matters, consisting mainly of crustaceans. Juveniles grow at an average monthly rate of 10 mm in Chilka lake, 14.4 mm in Ennur estuary, 16 mm at Adyar estuary and 24 mm in Covelong back waters. The adults show a growth of 5.6 and 7.0 mm in Males and Females at Tamil Nadu. Males and Females attain a length of 156 and 138 mm at the end of first year and 189 and 181 mm at the end of second year of life. Fishery is supported by the 0-year old in the estuaries and by 0-year (80-120 mm) and 1 year old (95–175 mm) in the marine region. Within the size range the modal sizes vary from place to place and season to season. The estimated annual total mortality in the fishery at one centre in south west coast is 3.1 in males and 2.1 in females.

Penaeus monodon are commonly distributed in the north-east coast. Number of eggs varies from 3 to 7 lakhs. Breeds in the same grounds as *P. indicus*. Foods consists of large crustaceans, vegetable matters, polychaetes, molluscs and fish. Largest recorded size is 337 mm. In the Chilka lake the juveniles grow at a rapid rate of 25 mm per month and at Tamil Nadu it reaches 160–170 mm size in 6 months in brackish water. Commercial catches are formed by 0–year and 1–year class. The species attains about 250 mm in one year.

Penaeus semisulcatus is more common on the east coast. The size of female at first maturity is 23 mm carapace length. Fecundity ranges from 67900 to 660900 eggs in different sizes. June to September and

January, February are peak spawning season in Gulf of Mannar and Palk Bay. Maximum size attained by this species is about 250 mm. It consumes large quantities of animal matter as well as diatoms and algal filaments. In the estuary it grows to about 150 mm forming the 0-year class. The marine fishery is contributed by sizes ranging from 120 to 230 mm consisting of both 0-year and 1-year classes.

Metapenaeus monoceros is distributed in both estuarine and marine regions. The species attain maturity in the sea after 120 mm size. Its fecundity ranges from 155000 to 338000 eggs. Peak spawning in July- August and November–December. It grows to a maximum size of 190 mm. Feeds mostly on small crustaceans. In Godavary estuary migration out of the estuary is mostly nocturnal and immigration mostly at dawn. The estuarine fishery is contributed by 0-year class. Marine fishery mostly contributed by sizes 125 mm to 150 mm of the 1-year class.

Distributed in the northern region of the coast, *Metapenaeus brevicornis* attains a maximum size of 135 mm and maturity at about 75 mm. In the Hoogly estuarine system there are two spawning seasons, in March-April and July-August, Major food items are vegetable matters and erustacean remains. Growth rate varies with salinity and temperature of the environment. In the estuary the sizes range from 15 to 115 mm constituted by 0 to 2 year groups.

Metapenaeus dobsoni is distributed upto a depth of about 40m, with large quantities in the brackish water areas. The species attain its first maturity at 64 mm length. Fecundity ranges from 34500 eggs in 70 mm prawn and 160000 eggs in 120 mm size. It attains a maximum size of 130 mm and breeds in the inshore waters inside the 25 m depth region. The species is a detritus feeder. Juveniles grow in the estuarine environment at an average monthly rate of 10 mm. Bulk of the fishery in the backwaters and the sea is supported by 3-12 month old prawns. The total annual instantaneous mortality rate on the west coast ranges from 3.1 to 3.8.

There are several other species of penaeid prawns which occur in small quantities in the fishery of different areas at different seasons. The important non-penaeid species which contribute to the fishery of mostly Andhra coast and in the northern region are *Acetes indicus, Palaemon tenuipes, Palaemon styliferus* and *Hippolysmata ensirostris*.

Spiny lobsters are distributed along the south-east coast and forms a good fishery at Tuticorin, Mandapam areas and Tamil Nadu. The important species are *Panulirus homarus, P. ornatus, P. versicolor.* The first two species are equally abundant. Peak seasons are January to March and July to September. Along the Bengal coast, *P. polyphagus* is the dominant species. In the south-east coast the sizes of lobsters in the fishery ranges from 110 mm to 370 mm. In Mandapam area an estimated population of 2.6 tonnes of lobsters was obtained with a rate of exploitation of 22.7 per cent.

An exhaustive list of different species of prawns that are found in Indian seas, together with vernacular names, wherever available, coasts of India, where they occur and the largest size to which they grow are presented below. The different genera have been arranged in the order of their commercial importance.

Genus–*Penaeus*

(1) *Penaeus indicus*: Chapra, Available in all coastal waters, estuaries and backwaters Maximum length 20–23 cm.

(2) *Penaeus monodon*: Tiger prawn, More on east coast, especially in Orissa and Bengal. On the west coast more on the northern sector. Largest Indian marine prawn, Attain a maximum size of 30–32 cm.

(3) *Penaeus merguiensis*: Occur in the middle regions of east and west coasts. Sparingly in other regions. Maximum length observed 24 cm.

(4) *Penaeus semisulcatus*: More common on east coast. Attain a maximum length of 23–25 cm.

(5) *Penaeus penicillatus*: Occur in coastal waters of north Bombay and Orissa coasts. Attain a maximum length of 21 cm.

(6) *Penaeus japonicus*: Sparingly occurs in Tamil Nadu (Pulicat lake) and Bombay coasts. Attain a maximum length of 27 cm.

(7) *Penaeus canaliculatus*: Very sparingly occur on south west coast. Grows to a maximum length of 15 cm.

(8) *Penaeus latisulcatus*: Only one or two specimens obtained from south-west coast.

Genus–*Metapenaeus*

(9) *Metapenaeus dobsoni*: Largest single species landed in India. Found in all coastal waters. More common on south west coast. Attain a maximum length of 12.5 cm.

(10) *Metapenaeus affinis*: Occur all along Indian coast. Attain a maximum length of 18 cm.

(11) *Metapenaeus monoceros*: Available along entire coast line, estuaries and backwaters. Largest size of 18 cm was recorded.

(12) *Metapenaeus brevicornis*: Available only in the northern region of both east and west coasts including estuaries. Attain a maximum length of 12.5 cm.

(13) *Metapenaeus ensis*: Reported only from east coast off Waltair. Maximum size attained 17 cm.

(14) *Metapenaeus lysianassa*: The species occur mostly on east coast off river Hoogly, Orissa coast and Gulf of Mannar. Rarely found on south west coast. Maximum length attained is 9 cm.

(15) *Metapenaeus burkenroadi*: Recently reported in Indian inshore waters and estuary in Cochin. Attain a maximum length of 10 cm.

(16) *Metapenaeus stebbingi*: Recently reported in India from Gulf of Kutch. Largest size of 12 cm was recorded.

(17) *Metapenaeus kutchensis*: The species form a good fishery in Gulf of Kutch area. Attain a maximum length of 13-14 cm.

(18) *Metapenaeus alcocki*: Recently reported from Gulf of Kutch area. Grows to a maximum length of 9.7 cm.

Genus–*Parapenaeopsis*

(19) *Parapenaeopsis stylifera*: Entirely marine species. Available all along the coast line especially west and south east coasts. Attain a maximum length of 14 cm.

(20) *Parapenaeopsis hardwickii*: Forms fairly a good fishery in north west coast, Bombay waters and off Godavari estuary. Maximum length recorded is 12 cm.

(21) *Parapenaeopsis sculptilis*: The species occur in northern regions of both east and west coasts only. The largest size is 16.5 cm.

(22) *Parapenaeopsis uncta*: The species was recorded from Orissa and south west coasts. Attain a maximum length of 10 cm.

(23) *Parapenaeopsis nana*: Recorded only from east coast, off Orissa and Tamil Nadu. Maximum length observed 5.5 cm.

(24) *Parapenaeopsis cornuta maxillipedo*: Available in Bombay, Kerala and Tamilnadu coasts. Not significant. Attain a maximum length of 12.5 cm.

(25) *Parapenaeopsis acclivirostris*: Occur both in east and west (Bombay) coasts. Maximum size available is 5 cm.

(26) *Parapenaeopsis tinella*: Only recently collected from east coast, Gulf of Mannar and Palk Bay. Maximum size is 5 cm.

Genus–*Macrobrachium*

(27) *Macrobrachium rosenbergii*: Available in most of the lakes and estuaries along the coast line of India. They grow to a maximum length of 30-32 cm.

(28) *Macrobrachium malcomsonii*: Most common in peninsular rivers draining into Bay of Bengal. Also available in river Indus. They attain a maximum length of 20-23 cm.

(29) *Macrobrachium villosimanus*: Available in Calcutta and Chittagong region, Maximum size attained is 15 cm.

(30) *Macrobrachium lamarrei*: Found in northern region of east coast, Chilka and Bengal.

(31) *Macrobrachium rude*: Found in south west region of west coast and throughout the east coast, especially deltaic Bengal, Orissa and Andhra coasts. They grow to a maximum size of 12-13 cm.

(32) *Macrobrachium idae*: More common along east coast. Attain maximum length of 10-11 cm.

(33) *Macrobrachium equidens*: Found only in Kerala in small numbers. Maximum length is 10 cm.

(34) *Macrobrachium mirabile*: Available mostly in Gangetic delta area. Grows to a maximum size of 6.5 cm.

(35) *Macrobrachium javanicum*: Occurrence restricted to deltaic Bengal. Attain a maximum length of 10 cm.

(36) *Macrobrachium sabriculum*: Occurs in deltaic Bengal, Chilka lake, Kerala and Tamil Nadu. Maximum size is 10 cm.

Besides, 24 others species of this genus have been reported to occur in India; but they are only of very minor importance.

Genus–*Palaemon*

(37) *Palaemon tenuipes*: Mostly found in the northern area both on east and west coasts and form a good fishery. They attain a maximum size of 8 cm.

(38) *Palaemon styliferus*: Occur in the northern area both on east and west coasts. Maximum size attained 9 cm.

These and the genus *Macrobrachium* belong to the family Palaemonidae.

Genus–*Acetes*

(39) *Acetes indicus*: The species is most common in Bombay waters, where they form an important fishery. They are also available on entire east coast both in sea and brackish waters. Attain a maximum size of 4 cm.

(40) *Acetes erythraeus*: Fairly good quantities are available in Bengal, Orissa, Tamil Nadu and Trivandrum coasts. They grow to a size of 2 to 3 cm.

(41) *Acetes sibogae*: Only available off Quilon, They grow to 3.5 cm in length.

(42) *Acetes serrulatus*: Available in Travancore coast. Attain 2 cm in length.

(43) *Acetes japonicus*: Available in west and south west coasts. Attain a length of 2.6 cm.

(44) *Acetes cochinensis*: Reported recently from Cochin waters. Attain 2 cm in length.

Genus–*Pandalidae*

Prawns of this family (pink in colour) have been found recently in deep water trawling operations off southern Kerala coast of 150 to 200 fathoms depth.

(45) *Parapandalus spinipes*: Found only in Arabian Sea off south west coast. Attain a size of 13 cm.

(46) *Plesionika martia*: Available in Arabian Sea and Bay of Bengal. Largest size attain is 12.5 cm.

(47) *Plesionika ensis*: Found only in Arabian Sea off south west coast. Attain a length of 12.5 cm.

Genus–*Heterocarpus*

(48) *Heterocarpus gibbosus*: The species is available in Arabian Sea, Bay of Bengal and Andaman Sea. Grows to a length of 14 cm.

(49) *Heterocarpus woodmasoni*: Available in Andaman Sea and Arabian Sea off Kerala coast. Attain a length of 13 cm.

Genus–*Hippolysmata*

(50) *Hippolysmata ensirostris*: The species forms a fairly good fishery in Bombay coast and Godavary estuary. Present in most regions in small numbers. Attain a length of 8 cm.

(51) *Hippolysmata vittata*: Present in small numbers in India Seas. Attain a length of 4 cm.

Genus–*Trachypenaeus*

(52) *Trachypenaeus pescadorensis*: Recently reported from south east and south west coasts. Attain a length of 9 cm.

(53) *Trachypenaeus sedili*: Recently reported from Visakhapatnam coast. Attain 6 cm in length.

(54) *Trachypenaeus curvirostris*: Available both in east and west coasts in small numbers. Attain 9.5 cm in length.

Genus–*Atypopenaeus*

(55) *Atypopenaeus stenodactylus*: Available both in east and west coasts off Tamilnadu and Bombay. Attain a length of 5 cm.

Genus–*Metapenaeopsis*

(56) *Metapenaeopsis stridulans*: Available in Bombay as well as northern region of east coast. Attain a length of 10 cm.

(57) *Metapenaeopsis mogiensis*: Occur off Malabar coast and Andaman Islands. Attain a maximum size of 9 cm.

(58) *Metapenaeopsis andamanensis*: Occur in deep waters of south west coast and Andamans. Recently caught at 150-200

fathoms off southern Kerala. Attain a maximum size of 13.5 cm.

(59) *Metapenaeopsis philippii*: Recently caught from deep waters of south west coast. Attain a maximum length of 13 cm.

Genus–*Parapenaeus*

(60) *Parapenaeus longipes*: Found in sea, off Mangalore, Cochin, Ganjam, Visakhapatnam and mouth of Hoogly river. Grows to a maximum size of 8 cm.

(61) *Parapenaeus fissurus*: Only available in east coast off Ganjam and Andamans, Attain 12 cm in size.

(62) *Parapenaeus investigatoris*: Occur in Gulf of Mannar, Pulicut lake, Andaman sea and off Cochin. Attain a maximum size of 8 cm.

Genus–*Penaeopsis*

(63) *Penaeopsis rectacuta*: Occur in seas off Tamil Nadu, Andamans and deeper waters off Kerala coast. Attain a maximum length of 13 cm.

Genus–*Solenocera*

(64) *Solenocera indica*: Available both in east and west coasts. Attain a length of 11.4 cm.

(65) *Solenocera pectinata*: Occur off south-west coast. They are too small.

(66) *Solenocera hextii*: Available along the entire west coast and Bay of Bengal

(67) *Solenocera choprai*: Occur only in Arabian Sea.

(68) *Solenocera koelbeli*: Available only in south west coast.

(69) *Solenocera melantho*: Recently reported from Godavari estuarine system.

Genus–*Hymenopenaeus*

(70) *Hymenopenaeus aequalis*: Available off south west coast and Andamans.

Genus–*Sicyonia*

(71) *Sicyonia lancifer*: Occur in Gulf of Mannar and south west coast.

Genus–*Aristeus*

(72) *Aristeus semidentatus*: Recently caught at 150–200 fathoms depth off Cochin and Alleppey. Largest prawns caught in these operations.

(73) *Aristeus alcocki*: Found in Bay of Bengal, Arabian sea, Laccadives and Cape Comorin.

(74) *Aristeus virilis*: Found in Andaman sea.

(75) *Aristaemorpha woodmasoni*: Available in Bay of Bengal, Andaman sea and Arabian sea.

Malaysia

Prawn fishing is traditional along the west coast of Peninsular Malaysia, where the main traditional fishing gears have been bag nets, seines and gill nets. With the introduction of trawling in mid-sixties and its subsequent expansion, the otter trawl is now the major gear employed in the prawn fishery. Trawl fishing accounted for about 76 per cent of total prawn landings on the west coast. Six genera consisting of 28 species of penaeid prawns from the commercial catches have been recorded. These are the, *Metapenaeus, Parapenaeopsis, Penaeus, Trachypenaeus, Metapenaeopsis* and *Solencera* of which *Metapenaeus, Parapenaeopsis* and *Penaeus* are the major genera of commercial importance. The most important species are *Metapenaeus affinis, Metapenaeus brevicornis, Metapenaeus lysianassa, Parapenaeopsis coromandelica, Parapenaeopsis hardwickii, Penaeus merguiensis* and *Penaeus monodon*. The most important genus commercially is *Metapenaeus* due to its abundant occurrence in commercial catches, while the genus *Penaeus* although not abundant, command a high price by virtue of the quality of their meat and large size.

Sergestid shrimp (*Acetes* sp) is another important species commercially available in the west coast of Peninsular Malaysia.

Landing of prawns on the east coast of Peninsular Malaysia depend partly on coastal semi-resident stocks which show seasonal fluctuations and partly on migrant stocks which appear in the coastal waters during the period of the north-east monsoon.

The penaeid prawn species recorded from the commercial catches include all the 6 genera of penaeid prawns found on the west coast of Peninsular Malaysia. However three genera appear to be found only on the southern part of the East Coast i.e. East Johore State. These are the genus *Solenocera, Metapenaeopsis* and *Trachypenaeus.*

Sri Lanka

The narrow coastal stretch extending from Colombo to Chilaw at depths of 6 to 20 m is a prawn ground that has been used for a long period of time. Prawn resources were located at the north of the island, a new fishery ground between Kachehativu and Rameswaram temple, and a narrow stretch of prawn ground from Pesalai at 6-10 m depth extending along the Palk Strait upto Dhanuskodi Point. A new resource was also located south-east of the Mullaithivu Light House at depths 16–24 m stretching southwards to a point west of Pullimoddai.

Mature females of small prawns such as *Metapenaeus dobsoni* and *Parapenaeopsis stylifera* are found throughout the year. *Penaeus indicus* breeds in February, March, April and again in November and December. Most of the females of *P. semisulcatus* are caught in November and most of them are mature.

In *P. indicus* size varies upto 200 mm. Below 135 mm, 90 per cent females are not developed. Between 135-185 mm gravid, developed and developing females are found. In *P. merguiensis* size varies upto 255 mm, gravid, developed and developing females are found. Size varies upto 205 mm in *P. semisulcatus*. Below 90 per cent females are not developed. Between 140-205 mm gravid, developed and developing females are found. Beyond 200 mm *P. monodon* though in few, gravid females have been found.

In Negombo, over 50 per cent of the females of *P. merguiensis* were seen to be gravid during the months of October to February, the peak period being in February. In Chilaw it is during December and January. In the same place 50 per cent of the females of *P. semisulcatus* are seen to be gravid during January and February, the peak being in January. In Negombo nearly 50 per cent of the females of *P. indicus* are seen to be gravid in February. In Chilaw, over 50 per cent of the females of *P. indicus* are seen to be gravid during January and February, the peak being in February. Specimens of gravid *P. monodon* females are present only during the months of November, December, January and February. Its breeding cycle may therefore be similar to other species.

Thailand

In the west coast or Indian Ocean coast of Thailand which extends west ward to Andaman and Nicobar Islands, northward to Mayanmar waters and south ward to west coast of Peninsular Malaysia and Sumatra Island harbour penaeid prawns like *Penaeus semisulcatus*, *P. monodon* and *P. merguiensis* to form commercial fishery. *P. semisulcatus* attain first maturity at 160 mm length and spawn during January and February. *P. mondon* however is sexually matured when they are 200 mm in length and their breeding season is June–August and February in Thailand coasts. *P. merguiensis* attain sexual maturity at smaller size, 130 mm, and is found to breed during July–August in Thailand coasts.

Chapter 4
Occurrence of Marine Prawns in Indian Seas

The fishing industry, during the past has been allured into prawn trade due to high price of prawns and ready export market. This led to the catching of large quantities of unmanaged prawn stocks from the coastal waters along the west coast. These stocks fluctuated from one season to another during different years. After years of such exploitation when the total landings came down, doubts were raised that prawn stocks were being over exploited along Kerala coast. Studying the problem, opinions were put forth that the diminishing of the prawn catch might have been due to change in the environmental conditions or over exploitation or both the factors acted simultaneously. Besides in some areas, small and undersized prawns are caught in large quantities. This practice have had serious implications on natural prawn stock.

On the east coast the exploitation of prawns started much later and was concentrated off the coast of West Bengal, Orissa and Andhra Pradesh. These fishing grounds exist off the mouths of major estuaries and lakes like Hoogly-Matlah, Mahanadi, Chilka lake, Godavari and Pulicat lake. In early stage of life cycle prawns come to coastal areas, in the vicinity of estuaries and at this state they were caught in large numbers causing the decrease in catch year to year, till ventures were given up in many areas as it was uneconomical. The magnitude was so much that during 1977 around 30 tonnes of juvenile prawns were caught, processed and exported from Paradeep and Visakhapatnam. In United States the industry purchased these prawns at a price which varies around one dollar per kg and made extruded products. But by then the killing of undersized prawn adversely affected not only the natural prawn stocks but also the industry.

Even though a total of 99 species of marine prawns have been reported from Indian seas, those that are commercially important are less than one-fifth of this number. Only 14 species of penaeid and 4 species of non-penaeid prawns are of commercial importance.

The total annual landings of penaeid prawns in India (average of 10 years, 1959-1968) was 47538 tonnes, contributed by 14 major species. They are in order of the magnitude of their occurrence,

1.	*Metapenaeus dobsoni*	33.81 per cent
2.	*Parapenaeopsis stylifera*	17.17 per cent
3.	*Penaeus indicus*	11.68 per cent
4.	*Metapenaeus affinis*	9.66 per cent
5.	*Metapenaeus brevicornis*	6.89 per cent
6.	*Metapenaeus monoceros*	4.40 per cent
7.	*Parapenaeopsis hardwickii*	3.62 per cent
8.	*Penaeus monodon*	2.78 per cent
9.	*Parapenaeopsis sculptilis*	2.52 per cent
10.	*Solenocera indicus*	1.34 per cent
11.	*Penaeus merguiensis*	1.34 per cent
12.	*Penaeus semisulcatus*	1.00 per cent
13.	*Parapenaeopsis uncta*	0.26 per cent
14.	*Penaeus penicillatus*	0.18 per cent
	Unidentified	3.35 per cent

The total annual landings of non-penaeid prawns in India is however, 34161 tonnes (average of 10 years; 1959-1968). The contribution of different species are:

1.	*Acetes indicus*	– 39.56 percent
2.	*Palaemon tenuipes*	– 17.89 per cent

3.	*Palaemon styliferus*	--11.52 per cent
4.	*Hippolysmata ensirostris*	--7.02 per cent
	Acetes spp.	--2.93 per cent
	Palaemon spp.	--0.02 per cent
	Unidentified	--21.06 per cent

About 52 species of prawns and shrimps that are either commercially exploited at present or have great commercial potentials in the Indian waters. The prawn fishery consists of penaeid and non-penaeid prawns. The penaeid prawns which constitute about 62 percent of the total prawn landings are the commercially important ones. The fishery is supported by multiple species that coexists in the fishing grounds and is characterised by wide seasonal and annual fluctuation in abundance. Most of the penaeid prawns are subjected to exploitation in the juvenile phase. In terms of landings Maharashtra ranks first, followed by Kerala, Andhra Pradesh, Tamil Nadu, Gujrat, Karnataka and West Bengal and Orissa.

Until a few decades back the prawn fishery was concentrated by and large on the west coast of India. But exploratory survey revealed existence of commercially exploitable stocks of good sized prawns along the east coast. Exploratory surveys carried out along the continental shelf edge and slope of the south-west and south-east coasts have located potentially rich fishing grounds of deep sea prawns. These grounds on the south-west coast are about 5000 sq. km. in extent and the magnitude of the harvestable resource from this area has been estimated to be about 5300 tonnes per year.

The prawn fishery of India is both dynamic and complex. It is supported by multiple species that coexist in the same fishing ground. The commercial penaeid prawns are subjected to exploitation in the juvenile and adult phases of their life in two different ecosystems of estuaries / backwaters and sea. Wide fluctuations in the catch are observed in all the regions. Neverthless their biological features, such as, their large fecundity, protracted breeding season, faster rate of growth, short life span and their ability to withstand wide environmental changes help to maintain their population. However, the exploitation on prawn resources from the wild stock has, in recent years, increased so rapidly that in certain areas of our coast it has almost reached or has already reached the optimum level.

Occurrence and Prevalence of Commercial Prawn Species in Maritime States of India

Gujarat and North Maharashtra Coasts

Acetes indicus (October-December), *Palaemon tenuipes* (September–December), *Hyppolysmata ensirostris* (September–December), *Solenocera indica* (October–May), *Parapenaeopsis stylifera* (September–February), *P. hardwickii* (September–February), *P. sculptilis* (September–February) and *Metapenaeus affinis* (September–December).

South Maharashtra Coast

Parapenaeopsis stylifera (January–December), *Metapenaeus affinis* (November–December), *M. dobsoni* and *Penaeus merguiensis* (January–December).

Karnataka Coast

Metapenaeus dobsoni (May–September), *M. affinis* (March–May), *Parapenaeopsis stylifera* (March–May) and *Penaeus indicus* (April–May, November)

Kerala Coast

Metapenaeus dobsoni (July, October–November), *M. affinis* (March–April), *M. monoceros* (January), *Parapenaeopsis stylifera* (September–December) and *Penaeus indicus* (March–April)

Tamil Nadu, South West Coast

Penaeus indicus (May–November)

Tamil Nadu, South East Coast

Penaeus indicus (May–July), *P. monodon* (May–July), *P. semisulcatus* (October–January, May–July), *Metapenaeus monoceros* (May–July) and *M. dobsoni* (May–July).

Andhra Pradesh Coast

Metapenaeus monoceros, M. dobsoni, M. brevicornis, Penaeus indicus and *P. monodon*

South Orissa Coast

Penaeus monodon (April–August), *P. indicus* (January–February and June–August). *Metapenaeus dobsoni* (January–February, October) *M. monoceros* (April, July and November) and *M. brevicornis* (August–December)

There are fourteen species of penaeid prawns landed from Puri coast, Orissa, during 2006. Out of the 1444 tonnes *Penaeus penicillatus* contributed 716 tonnes (50 per cent of the total prawn in Puri district), followed by 176 tonnes by *Metapenaeopsis stridulans*, 126 tonnes of *Parapenaeopsis hardwickii* and 102 tonnes by *Metapenaeus affinis*. *P. merguiensis* and *P. penicillatus* were mainly caught by bottom set gill nets, where as, all the other varieties were the main stay of trawl nets. In almost all months, juveniles of penaeid prawns like *Parapenaeus longipes*, *M. stridulans* and non-penaeid prawns like *Nemato-palaemon tenuipes* were found in trawl catches.

Table 3

Name of Species	Landings (Tonnes)		
	Trawl Net	Bottom-set Gill Net	Gill-Net
Metapenaeus affinis	88.772	9.223	3.929
M. dobsoni	87.111	0	0.491
M. lysianassa	0.280	0	0
M. monoceros	41.593	0	0
Metapenaeopsis stridulans	175.857	0	0
Parapenaeopsis hardwickii	126.348	0	0
Penaeus indicus	9.015	0	0
P. japonicus	23.362	0	0
Parapenaeus longipes	37.532	0	0
P. merguiensis	0.424	5.516	1.586
P. monodon	12.378	2.170	0
P. penicillatus	16.299	700.136	0
P. stylifera	31.505	0	0
Solenocera crassicornis	66.794	0	0
Others	0.285	4.014	0
Total	717.555	721.059	6.006

North Orissa and West Bengal coast–*Penaeus indicus* (October–December), *P. merguiensis* (October–December), *P. monodon* (October–December) and *Metapenaeus affinis* (October–December).

The abundance of tigers gradually increased from 11-60 m and then gradually declined beyond this depth. The proportion of tigers in the total prawn catches also indicated a similar trend. The abundance of white was better in 11-40 m depth and then declined gradually beyond this zone. The proportion of whites gradually declined from 11-20 m to 91-100 m. The abundance of browns gradually increased from 11-20 m to 91-100 m. Similarly the proportion of browns in the prawn catches also increased gradually from 11-20 m to 91-100 m.

The abundance of total prawns was more in November and December in 11-20 m depth range, September–December in 21-30 m, May–November in 31-40m, May to October in 41-50m, May to November in 51-60 m, August–December in 61-70 m, July to January 71–80 m, August–January 81–90 m and August to December in 91–100 m depth ranges. In general the abundance was less in February to April in all the depths.

The abundance of tigers was more in October in 11-20 m, August to September in 21-30 m, May to October in 31-40 m, July to March in 41-50 m, May–August and December–February in 51–60 m, January, August and October in 61-70 m and July to December in 71-80 m depth ranges.

The abundance of browns was more in August–October in 21-30 m, May–June in 31–40 m, April–July in 41–50 m and May–July in 51–60 m depth ranges. In the depth range of 21–60 m 'browns' were abundant in almost all the months.

Marine Prawns

The penaeid prawns belonging to the genera *Penaeus* and *Metapenaeus* spawn in the sea but the post larvae (1-2 cm in length) enter the estuaries and backwater areas in large numbers and grow rapidly. These areas serve as natural nurseries for the juveniles. The euryhaline nature of these prawns enables them to colonize the estuaries and backwaters. The smallest and largest size of prawns collected from brackish waters are 15-25 mm and 80 mm respectively. Reverse migration into the sea take place soon after monsoon rains, the prawns being carried in the current flowing into the sea. Spawners do not leave the sea afterwards to their former habitat.

Penaeus canaliculatus, characterized by two flagella in antennules, 1 to 3 chelate walking legs and the pleurae of second abdominal segment overlap those of first segment. Rostrum is serrated on both margins. One tooth on ventral margin of rostrum. Body with transverse coloured bands. The species is found in both the coasts of India and are more common in creeks.

Penaeus semisulcatus is found in both east and west coasts of India. It support commercial fisheries in the Sunderbans and along the east coast. The species is characterized by two flagella in antennules, 1 to 3 chelate walking legs and the plurae of second abdominal segment overlap those of first segment. Rostrum is serrated on both margins; three teeth on ventral margin of rostrum. Its adrostral groove is deep, extending upto half-way along the carapace. Exopodite present on fifth leg with oblique hepatic ridge.

Penaeus monodon is characterized by two flagella in antennules, 1 to 3 chelate walking legs and the pleurae of second abdominal segment overlap those of first segment. Rostrum is serrated on both margins. Three teeth is present on the ventral margin. Adrostral groove is shallow and does not extend beyond the rostrum. Exopodite on fifth leg is absent, hepatic ridge is longitudinal. The species is more common on the east coast of India, especially in Bengal and Orissa where it contributes to a fishery. Small numbers are found all along the Indian coast.

Penaeus monodon is distributed in the east and west coast of India and form a large scale fishery in the bheries of West Bengal, Chilka lake in Orissa and coastal Andhra Pradesh. The species is omnivorous with preference for animal matters like polychaetes, crustaceans, insects small molluscs etc. The rate of growth is 17.55–55.20 mm per month under short term crop. 40-50 mm juveniles attain 130-140 mm in two months in a well prepared ponds. Grows to 180-250 mm in one year. Maximum recorded size of the species is 320 mm. The species do attain maturity in ponds but breeds in the sea near to the estuarine mouths, and post-larvae of 10-20 mm size migrate to the estuaries, lakes and backwaters.

Penaeus monodon, although common is comparatively less abundant than *P. semisulcatus.* Primary peak was from January to

March in all the biotopes. The secondary peak was noted in September in the estuary, August, September in backwater and November, December in the mangrove region. *P. monodon* was absent in May, June and July months in the mangroves. The maximum length recorded was 147 mm.

Penaeus indicus is found all along the coasts, more abundant in the southwest and south east coasts. Form a considerable fishery in the estuaries and backwaters. The species is characterized by two flagella in antennules, 1 to 3 chelate walking legs and the pleurae of second abdominal segment overlap those of first segment. Its rostrum is serrated on both margins. More than three teeth (4 to 6) are present on the ventral margin of the rostrum. Body is without transverse coloured bands. Gastro-orbital carina occupying the posterior 2/3 distance between hepatic spine and orbital angle. Rostral crest only feebly elevated, not triangular in profile.

Penaeus indicus is available both in east and west coast of India. The species is more abundant in the coastal zones of Karnataka, Kerala, Tamil Nadu, Andhra Pradesh, Orissa and West Bengal. It is a omnivorous feeder. Food items are mainly detritus, small crustaceans, polychaetes etc. Attains a marketable size of 80 to 120 mm in 90-100 days under cultural conditions. The species do not attain maturity in culture condition. It attains maturity in marine environment and breeds in the deeper waters of the seas. Hatching of fertilized eggs takes place in the sea. Post-larvae migrate to estuaries and backwaters in brackish water environment. Maximum recorded size is 230 mm.

Penaeus merguiensis, through found throughout the Indian coast, support commercial fisheries only in Goa, Karwar, Visakhapatnam and Puri. The species can be identified by the presence of two flagella in antennules, 1 to 3 chelate walking legs and overlapping pleurae of the second abdominal segment on the first segment. Its rostrum is serrated on both margins. More than three teeth (4 to 6) are present on the ventral margin of the rostrum. Body is without transverse coloured bands. Gastro–orbital carina occupying the middle 1/3 distance between hepatic spine and orbital angle. Rostral crest conspicuously high, forming an elevated triangular crest.

P. merguiensis showed a decreasing trend of abundance from January onward and recovered to reach its peak in May. It showed a decrease from June to December in the estuary. The size of the population becomes smaller from August and tapers off. They were absent in November and December in back water. In mangrove region, *P. merguiensis* was abundant in January and February but was less in the remaining months. The species was absent in March, April and September to December. More number of this species were found to occur in the backwater, than in the other two biotopes. 138 mm was the maximum length recorded in the biotopes.

Parapenaeopsis stylifera, characterized by the presence of exopodite on the base of fifth leg, rostrum with serrated dorsal margin, hepatic carina running on to branchiostegal tooth, strong telsonic spines and dull brown in colour is found all along the coast. The species is commercially exploited in the west coast. The species at no stage enter in the brackish water and is truly marine in nature.

Parapenaeopsis sculptilis though found all along the Indian coasts, but is more common in Bombay and estuaries of West Bengal. The species is characterized by the presence of exopodite on the base of the fifth leg, serrated margin on the dorsal side of the rostrum, weak telsonic spines, if present and with dark transverse bands on the body. Its hepatic carina does not reach banchiostegal tooth.

Metapenaeus monoceros occur throughout the Indian coasts. Juveniles abound in estuaries. Mature large prawns are found only in offshore seas. Rostrum is straight without rostral crest. The first abdominal segment is carinated. Rostrum extends beyond the second segment of the antennular peduncle. Exopodite absent on fifth leg.

Metapenaeus monoceros is available in the coasts and estuaries of Kerala, Tamil Nadu, Andhra Pradesh, Orissa and West Bengal. It feeds on different types of animal and vegetable matter. The species grows well in brackish water ponds. Attain 100-130 mm in about 6 months. Maximum recorded length of the species is 180 mm. Maturation and breeding of the species occur in sea. Males attain maturity in ponds and stray females of advanced maturity stages are encountered in culture conditions.

Metapenaeus monoceros is abundant from January to April and August to November in the estuary, backwater and mangrove biotopes. It was caught in good numbers from the backwaters. The major peak was observed during June to September and a minor peak in December. The primary peak was observed in September and October in both estuary and mangrove regions. A minor peak was noticed during April in the estuary and during January–

February in the mangroves. The maximum size observed was 81 mm.

Metapenaeus brevicornis is characterized by straight rostrum with rostral crest. The first abdominal segment is not carinated. Rostrum does not extend beyond the second segment of the antennular peduncle. Exopodite is absent on fifth leg. The species is found only in northern regions of both the coasts, where the juveniles occur in estuarine areas and adults in the inshore waters.

Metapenaeus brevicornis occur abundantly from January to April and during September and October. It occur although not in abundance from May to August and November in the estuary. In the backwater the primary peak is during January–February and second peak during August–September. In the mangrove region, two major peaks of abundance with a minor peak were observed in February, June and August respectively. The maximum size was 70 mm.

Metapenaeus brevicornis is available in the northern zones of east coast, West Bengal, coastal Andhra Pradesh and Orissa. Forms a dominant fishery in the estuaries of West Bengal. The species in omnivorous feeding on detritus, animal and vegetable matter. It grows to 80-100 mm in 6 months under culture conditions. Maximum recorded size is 125 mm. The species mature and breed in the sea. Males attain functional maturity in brackish water ponds and impregnates immature females. Females of advanced maturity are not uncommon in the ponds.

Metapenaeus affinis is commercially important all along the Indian coasts. The species is characterized by slightly curved rostrum without noticeable crest. Its last pair of thoracic legs surpass the antennal scale. Exopodite is absent on fifth leg.

Metapenaeus affinis forms good fishery in the coasts of Gujarat, Maharashtra, Karnataka and Kerala. Though omnivorous, it prefer vegetable matter. A monthly growth rate of 8.5 mm was recorded from estuarine environments. Maximum recorded size is 180 mm. Maturation and breeding of the species takes place in the inshore areas of the sea.

Metapenaeus dobsoni is one of the most abundant species, occur, particularly along the Kerala coast. Adults inhabit in the inshore areas and juveniles frequently enter estuaries and backwaters. Rostral curvature of the species is pronounced with a crest. Last pair of thoracic legs falls short of middle of antennal scale. Exopodite is absent on fifth leg.

Metapenaeus dobsoni forms a good fishery in the west coasts of South Maharashtra, Karnataka, Kerala and Tamil Nadu. In the east coast the species is available upto South Orissa. The species feeds on foraminifers, amphipods, cladocera and diatoms. Maximum recorded size of the species is 125 mm. In the paddy fields of Kerala it grows to 61-63 mm in about 4 to 5 months. The species attain maturity and breeds in the sea. Post-larvae of 4-5 mm size migrate to the estuaries.

M. dobsoni is available in good quantity from January to May, begin to diminish in numbers and again stage a recovery and attain the peak in December in the estuarine waters. In backwater they form a primary peak in July and secondary peak during December. Its highest number is observed in December in the mangrove region. The maximum size of this species was 68 mm.

Penaeus semisulcatus forms a small fishery in the east coast of Tamil Nadu and Andhra Pradesh and also in the bheri fisheries of West Bengal. Generally carnivorous it feeds on polychaetes, crustaceans, small fishes and vegetable matter. Maximum recorded size of the species is 250 mm. Attain maturity and breed in the sea. Its post-larvae migrate to estuaries, lakes and back waters.

Penaeus semisulcatus is one of the penaeid prawns enjoying wide distribution along the coastal waters of India. It contributes to a major fishery and dominates the commercial catches along the southeast coast. This being a species growing to a large size, is in good demand in the export industry and is increasingly exploited by mechanized boats.

Penaeid prawns in their juvenile stage assemble in shallow brackish waters and estuaries all along the coast of India. Considerable variation occur both in the degree to which brackish water environment the commercial penaeids utilise and in the distribution of the parent and juvenile populations in the brackish and marine environments. The non-entry of *Parapenaeopsis stylifera* into the brackish water environment may be cited as an example. Although juveniles of *P. semisulcatus* are found in estuaries in other areas, an exception to the set pattern of life history of the species is noticed in the fishery observed in the shallow coastal waters of Tuticorin, which provide a purely marine environment.

The fishery of *P. semisulcatus* is limited to the shallow coastal waters extending to about 20 km from Pattanamarudur to Tuticorin. It is also exploited occasionally at Harbour-Point, Hare Island and Pattanamarudur. The substratum of the fishing ground is sandy, with corals and rocky patches. The ground slopes gently and is covered with a thick growth of marine plants, sea grasses and algae.

The inshore area is very shallow, the 6 metre contour being conspicuously broad from Vaippar to Tuticorin. It reaches the maximum width at Pattanamarudur and Tarravarculam to about 6 to 7 km away from the shore. No such broad and shallow fishing ground is noticed anywhere in the adjacent inshore sea. Most of the commercial penaeid prawns are known to be closely associated with shallow brackish water environment after their post-larval stages, showing a preference to shallow waters during their juvenile stages. The extensive shallow water area, though not brackish, along the Tuticorin coast may be one of the favourable factors for the occurrence of a juvenile population of *P. semisulcatus* here.

One of the salient features of this environment is the abundance of marine plants in the fishing ground. Occurrence of the juveniles of *P. semisulcatus* in marine habitat with thick growth of aquatic plants have been reported elsewhere. The occurrence of the species in large numbers where there is a good growth of eel grass (*Zostera capensis*) and relatively scarce when the growth of eel grass is poor has been observed in Durban Bay, on the east coast of Africa. The juveniles of *P. semisulcatus* measuring 3.2 to 17.0 mm in carapace length spent their life from late August to middle of October in the areas of Seto Island, Sea of Japan where *Zostera marina* is growing. After middle of October the species migrate the offshore areas with muddy bottom. The abundance of juveniles of *P. semisulcatus* in the shallow inshore waters off Tinnevelly coast rich in marine plants and their migration as they grow in size to deeper waters with muddy substratum indicate that its juvenile stage always prefers marine environment with thick growth of aquatic weeds. The species has been observed to form significant portions of the prawn catches of "Bheris" of West Bengal and are often well represented in the brackish water fishery of the west coat of India.

Production and Landings of Marine Prawns

Marine prawn fishery of India was confined to the traditional fishing sector till mid-nineteenth century. The production was, therefore, very low (85 thousand tonnes). The discovery of rich prawn grounds in the off shore waters, introduction of mechanized shrimp trawlers, modern processing techniques and high demand for processed prawns in the international market in subsequent years made an extra-rapid development in capture fisheries of marine prawns. At present prawns are intensively caught throughout the Indian coast upto 100 metres depth by a variety of fishing crafts, small, medium and large trawlers and the mechanized and non-mechanized country crafts, all trying hard to catch more prawn, often encroaching each other's fishing grounds.

The declining trend or stagnation of prawn catch from the conventional fishing grounds in many regions has led the prawn catchers to put increased efforts, by the way of moving to distant fishing grounds spending more time on voyage fishing, operation of ring seines in coastal waters and introduction of "trammel" net in place of bottom set gill nets to increase the catch returns.

Prawn fishing extends almost throughout the year on most part of the Indian coasts except for a short period of off-season during monsoon period. However, in Kerala peak fishing of prawn takes place during the southwest monsoon period especially by trawlers at Cochin and Quilon.

In the total marine fish production of India, prawns contribute 10-12 per cent of which 80-85 per cent are obtained from the west coast. About 80 per cent of the prawn catch is obtained from Maharashtra, Gujrat and Kerala alone. Tamil Nadu, Andhra Pradesh and Karnataka together accounts for 15 per cent and the remain 5 per cent are contributed by other maritime states and Union Territories. An estimated prawn landings of 85 thousand tonnes in 1958 was increased to 389 thousand tonnes, a four fold increase over the years. Between 1958 and 1967 the marine prawn production remain below 100 thousand tonnes per year and thereafter increased steadily till 1973 crossing the prawn landings by 200 thousand tonnes. There was a sudden fall in the landing in 1974 only to shoot up by 220 thousand tonnes in 1975 followed by a wide annual fluctuations till 1989. A higher production of more than 250 thousand tonnes have been obtained in subsequent years, followed by a record landings of 389 thousand tonnes in 1998.

The variations in prawn catch in different regions of Indian coast is due to the differences in the extent of continental shelf, bottom conditions, hydrographic features, connected estuarine areas and regional climatic conditions, which have a tremendous influence in the distribution and abundance of prawn stocks.

During 1978 trawl catch of prawn accounted for 51 per cent and the rest 49 per cent was harvested by the traditional fishing sector. The landings of marine prawns by traditional fishing sectors had diminished considerably over a period of 1980-1989, while trawl fisheries maintained increased landings in some of the regions. During 1985-89 period trawling contributed about 59 per cent and traditional fishing 41 per cent of total prawn landings. While Kerala, Maharashtra, Gujrat and Tamil Nadu contributed major portion of the trawl catches, Maharashtra and Gujrat contributed bulk of the prawn catch by the fixed bag net. Prawn catches were more in boat seines and shore seines in Kerala, Tamil Nadu and Andhra Pradesh. Prawns were landed by ring seine nets in Kerala and in the gill nets in Kerala and Andhra Pradesh under the traditional fisheries sector.

Several species of penaeid and non-penaeid prawn inhabits in Indian Ocean. Indian prawn fishery is therefore, multi-species in characters. The penaeid prawns account for on an average of about 65 per cent and non-penaeids 35 per cent in the total prawn fishery. While penaeid prawns constitute almost entire landings in Kerala, Tamil Nadu, Karnataka, Orissa, Goa and Pondicherry, the non-penaeids accounts for as much as 65 per cent in Gujrat, 38 per cent in Maharashtra, 45 per cent in West Bengal and 12 per cent in Andhra Pradesh besides the penaeid groups in the fishery.

92 per cent of penaeid prawns are caught by trawl nets. The important species that are caught in trawl nets are *Solenocera crassicornis, Metapenaeus kutchensis, Parapenaeopsis stylifera* and *P. hardwickii* along the Gujrat coast, *Metapenaeus affinis, M. monoceros, P. stylifera, P. hardwickii,* and *P. sculptilis* along the Maharashtra coast, *M. monoceros, M. affinis, Penaeus merguiensis* and *Metapenaeus dobsoni* along the Goa and North Karnataka coasts, *P. stylifera, M. dobsoni, P. indicus, M. monoceros* and *M. affinis* along the south-west coasts, *P. indicus, P. semisulcatus* and *M. dobsoni* along the south-east coasts, *M. monoceros, M. dobsoni, P. stylifera, M. brevicornis, P. indicus, P. monodon* and *M. affinis* along the Andhra coasts and *M. monoceros, P. indicus, P. monodon, P. merguiensis* and *P. hardwickii* along the north east coasts.

The traditional fishermen using their indegenous crafts catch both the penaeid and non-penaeid prawns with their traditional gears. The fixed bag nets (dol net) operating in the inshore waters of Gujrat and Maharashtra land the bulk of non-penaeid prawns (70-75 per cent) in Indian coast, of which about 75 per cent belongs to *Acetes* species. Among the penaeid prawns, caught by indigenous gears, the most dominant are *M. kutchensis, M. dobsoni, P. stylifera* and *P. indicus* along the southwest coast and *P. indicus, M. dobsoni, M. monoceros, M. brevicornis* and *P. monodon* along the east coast. Among the non-penaeids, besides *Acetes* species, *Nematopalaemon tenuipes* and *Exhippolysmata ensirostris* contribute bulk of the dol nets landings in Bombay–Saurashtra coasts and *Exopalaemon styliferus* on the West Bengal coast.

The offshore trawler operations and night fishing in search of new prawn fishing grounds have resulted in the occurrence of a number of new species in the prawn fishery, throughout the west coast and southeast coast. In Karnataka and Kerala coasts the landings of *M. monoceros* has increased considerably, since the trawlers have started catching prawn beyond the 50 m depth contour, where the species concentrate in huge numbers. In the voyage fishing trips at night the catch of *M. monceros* has also increased due to its nocturnal habits. Similar higher landings of *M. monoceros* and other associated species like, *P. canaliculatus, P. semisulcatus, Metapenaeus stridulans, Tarachypenaeus* spp have been observed in the night catchs of shrimp trawlers operating from Kerala coast. In the Bombay coast also, the catches of non-commercial species like *M. stridulans, Trachypenaeus curvirostris, Parapenaeus longipes* and *Solenocera choprai* have also increased as a result of fishing in deeper waters. Along the Tamil Nadu coast also a number of non–conventional prawn species have made their appearance in sizable quantities in trawl fishery due to extension of fishing into deeper seas.

Resources of Marine Prawns

Due to great demand in the international market and high priced commodity, prawn fishery has been given high priority in the fishing industry in past three decades all over the world including India. Penaeid prawn constituted 65 per cent and non-penaeids around 35 per cent. Prawn fishery landings from west coast of India has been always found to be on the higher side than the east coast. Many a times decline in the prawn landings has been recorded in some of the maritime states is due to the failure of post monsoon fishery and diversion of fishing activity in search of deep sea prawns. In Kerala, coast, *Parapenaeopsis stylifera* fishery has been reported to be always dominant almost every year, besides other penaeid species like *Metapenaeus dobsoni*. Peak spawning of *P. stylifera* is generally observed during May to December. Along east coast in Andhra Pradesh and Tamil Nadu major prawn species is of *Penaeus semisulcatus*, but a good fishery of *M. monoceros* particularly at Visakhapatnam has also been reported.

During 1999-2000 the crustaceans contributed 8.99 lakh tonnes forming 16.5 per cent of the total marine landings and showed a decline by 19.8 per cent ; over 1998-1999. The other reasons for decline in prawn catches are thought of due to;

Over fishing in some areas because of the introduction of large number of small mechanized boats in the inshore waters; increasing fishing efforts which has led to decrease in catch per unit of effort; and large number of boats engaged in fishing operations generally disturbs the shrimp beds resulting in migration of these shrimps in deeper waters.

The decrease of prawn population in the marine environment also occur when fishing effort reach beyond maximum sustainable yield (MSY) where fishing mortality and natural mortality together is more than recruitment and growth. Reclamation of backwater areas for the development and environmental pollution of coastal inshore areas have become vulnerable for survival of growing prawns. Juvenile prawns are generally caught in large numbers in the trawl nets by the fishermen due to lack of implementation of mesh size regulation. Even the brooders are caught and not allowed to spawn either during monsoon or post-monsoon season. They are caught live and sold to the owners of shrimp hatcheries at a very high cost. Each brooder when allowed to spawn in wild produces around 6-8 lakhs prawn seed.

Today trawls and purse seine nets are operated in a big way to catch prawns of the sea. The estimated penaeid prawn catch from major trawl fishery centres during 1999 were 2628t (4.1kg/hr) at Veraval, 11868t (7.4 kg/hr) at Mumbai, 227t (5.7 kg/hr) at Karwar, 2586t (1.7 kg/hr) at Mangalore- Malpe, 524t (26.8 kg/hr) at Kozhikode, 4168t (8.2 kg/hr) at Kochi and 7440t (5.2 kg/hr) at Sakthikulangara–Neendakara. In comparison to the landings of 1998 prawn fishery declined at all the above centres from 30 to 55 per cent. Major decline of 48.7 per cent occurred at Sakthikulangara with respect to *P. stylifera*. The species during 1999 was the main contributor to penaeid prawn fishery at Veraval (58 per cent), Mumbai (40 per cent), Karwar (54 per cent) Kocki (47 per cent) ad Sakthikulangara (47 per cent). *Metapenacus monoceros* formed 37 per cent prawn catch at Mangalore–Malpe. *Metapenaeus dobsoni* was the dominant species in the fishery at Kozhikode (47 per cent) and Kochi (48 per cent); while at Karwar it formed 34 per cent of the catch forming the second in importance. The species formed 14 per cent of prawn landings at Sakthikulangara and 12 per cent at Mangalore–Malpe. *Trachypenaeus* spp formed one of the main contributors to the fishery at Sakthikulangara and Mangalore forming 18.4 per cent and 19.6 per cent respectively. *Solenocera crassicornis* contributed to 29 per cent and 21 per cent to the prawn fishery at Veraval and Mumbai, thus becoming the second dominant species. At Mangalore–Malpe, *S. choprai* formed 14.5 per cent of the shrimp catch. *M. affinis* was one of the important component of shrimp fishery at Veraval (5 per cent), Mumbai (13 per cent), Karwar (10 per cent) and Kozhikode (11 per cent). *Penaeus indicus* at Kozhikode, Kochi and Sakthikulangara, *P. canaliculatus* at Mangalore–Malpe and Sakthikulangara and *P. semisulcatus* at Veraval, Mumbai and Sakthikulangara supported minor fisheries.

East coast contributed 33.7 per cent (55802 t) penaeid landings in the country during 1999-2000, a 4.8 per cent increase of 1998 (52253 t). In order of abundance, the statewise landings were 24965 t (44.7 per cent) in Andhra Pradesh, 23443 t (42 per cent) in Tamil Nadu, 4322 t (7.8 per cent) in Orissa, 2704 t (4.8 per cent) in West Bengal and 368 t (0.7 per cent) in Pondicherry. The prawn landings increased over previous year by 31 per cent in Andhra Pradesh, and 109 per cent in Orissa, but declined by 13 per cent in West Bengal, 17 per cent in Tamil Nadu and 48 per cent in Pudducherry.

Catch of penaeid prawn by trawling amounted to 176 t (12.5 kg/boat trip) at Tuticorin, 2263 t (2 kg/hr) at Chennai, 650 t (1.3 kg/hr) at Mandapam, 6745 t (8.5 kg/hr) at Visakhapatnam and 528 t (4.7 kg/hr) at Paradeep. Over the previous year the prawn landings increased by 232 per cent at Tuticorin and 8 per cent at Paradeep. The catch declined by 46 per cent at Chennai, 11 per cent at Mandapam and Visakhapatnam and 21 per cent at Kakinada. Along south east coast *Penaeus semisulcatus* was the dominant species at Tuticorin and Mandapam. *Metapenaeus dobsoni* supported 21 per cent of the landings at Chennai and 35 per cent at Kakinada. While *M. monoceros* contributed to 45 per cent of the fishery at Visakhapatnam *Parapenaeopsis hardwickii* accounted for 30 per cent at Paradeep. Other important species of the fishery were *Penaeus indicus* (17 per cent) and *Metapenaeopsis stridulans* at Chennai, *M. stridulans* and *Trachypenaeus pescadorensis* at Mandapam, *M. monoceros* (35 per cent) at Kakinada, *M. dobsoni* (16 per cent) at Visakhapatnam and *Parapenaeopsis stylifera* (12 per cent) and *Solenocera crassicornis* (7 per cent) at Paradeep.

Landings of non-penaeid prawn in India during 1999 a mounted to 154498 tonnes showing a decline of 11.2 per cent over 1998. The catch is inclusive of 9200 tonnes of pandalid prawns fished by deep sea trawlers from south west coast. Gujrat accounted for 58.2 per cent followed by Maharashtra (28.6 per cent). In comparison to previous year the fishery declined by 5.4 per cent in Gujrat and 34.3 per cent in Maharashtra.

Pandalid prawns like *Hetrocarpus woodmasoni* (41.2 per cent), *Plesionika spinipes* (31.5 per cent) *H. gibbosus* (9.3 per cent) and few others constitute deep sea prawn fishery along south west coast. A catch of 1.5 lakh tonnes of non-penaeid prawn exhibit a decrease by 11.2 per cent compared to previous year (1998). Gujrat contributed 58.2 per cent. *Acetes* spp while 80-99 per cent were landed at Mumbai, where in *Nematopalaemon tenuipes* formed 99.8 per cent.

Chapter 5

Life of Marine Prawns

The penaeid prawn is a coastal habitat family that changes its living grounds in the different stages of its life cycle. The growth stages and their corresponding living environment and life pattern changes for naturally raised prawn are given below:

Table 4: Life Stages of the Kuruma Pranw

Division	Beginning of Stage	Period	Life Style	Habitat
Embryo	Fertilization	0.6 day	Floating freely	Off-shore water
Larva	Hatching	30 days	Swimming freely	Off-shore water
Fry of gills	Development	15 days	Bottom life	Tideland
Juvenile	Body development stabilizes	2 months	Bottom life	Tideland → Beach
Pre-mature	Begins sexual development	9 months	Bottom life	Beach → Open sea (Bay water)
Mature	Sexual maturity completed	24 months	Bottom life	Open sea off shore

The egg laying and fertilization patterns of marine prawns can be divided into two main types:

(a) Dendro Branchiata Type

At the time when the berried female releases her eggs in the water, she also releases the sperm she received from the male in the mating process that precedes the spawning. Mating process takes place only between 0 AM to 3 AM and no other time. Prior to copulation the prawns creep about at the bottom, only one male following one female. In a few minutes the female moults becoming soft, the male remaing hard without moulting. After moulting the male advances and embraces the female on the ventral side and they swim with their bodies inclined. Copulation lasts for 3 or 4 minutes after which they seperate without caring for each other. During copulation the male ejects the spermatophores containing sperms through the petasma, protruding from the base of fifth pereopods of the male into the thelycum of the female. The spermatophores reach the seminal receptacle through the thelycum. The same male cannot copulate twice in the same night, since all the spermatophores are deposited in one female.

The fertilization and hatching of the eggs then takes place in the sea water. All penaeid prawn species belong to this type. This type is characterized by large number of eggs released by each female and the shortness of the larval period. A parent prawn (spawner) with a body length of 20 cm is said to lay between 250 and 800 thousand eggs.

(b) Incubation Type (Pleocyemata)

The female attaches the eggs she has laids to her belly, fertilizes and holds them protected there until they hatch. After hatching she then ejects them out into the sea water. This incubation type is seen in shrimps and prawns like *Caridea* and *Stenopodidea*. Two distinctive features of this type are that the number of eggs are fewer than Penaeids and the incubation period is longer. For example an Ise lobster (*Panulirus japonicus* of 12-13 cm body length) gives birth to 29000 eggs and the incubation period lasts a full 8–10 months. Among *Carides*, *Macrobrachium rosenbergii* lays 40000 to 100000 eggs and incubation lasts 20 days.

The first researcher in the world to undertake the study of artificial hatching methods and the early life cycle of *Penaeus japonicus* was Dr. Motosaku Hudinaga (1903-73) of Japan. Dr. Hudinaga first succeeded in the artificial hatching of marine prawn eggs in 1933.

At that time he was able to raise the young upto zoea stage. But as they reach their feeding stage of their life cycle they began to die one after another and the raising process ended in failure. In 1940 however, he tried giving pure cultured *Skeletonema*, a kind of diatom to the zoea and succeeded in getting them to feed for the first time. In 1955 he tried giving newly hatched brine shrimp young as feed and succeeded in raising the prawn from mysis stage to the post-larva stage. In this way after twenty years of research and experimentation he succeeded in developing the theory of artificial hatching and the methods of larvae rearing for the *Peneaus japonicus*.

The egg inside the ovary is covered with a layer of follicle cells and possesses 2 or 3 nucleoli. There is no vitelline membrane. When eggs grows it absorbs the follicle cells and in a mature egg the latter grow smaller becoming a thin membrane, the nucleoli also becoming smaller and increasing in number. After all the follicle cells are absorbed, a jelly-like substance is found on the surface of cytoplasm. The egg is an irregular sphere before fertilization discharging radially the whitish, transluscent, jelly-like substance from within. The latter becomes granular and seperates from the eggs seven minutes after spawning, but a little may remain even after the formation of the fertilization membrane. By the time the first segmentation is over the jelly may completely disappear and the eggs become spherical.

During fertilization a few spermatozoa, even upto ten, reach the egg surface one minute after spawning, and after the disappearance of the jelly mostly. At the place of contact of the spermatozoa on the egg, entrance cone appear according to the number of spermatozoa that have reached the egg. The cytoplasm of a cone protrudes upto the head of the spermatozoon and into this the head sinks. Soon the cytoplasm shrinks and drags the spermatozoon into the egg through the egg surface. Only one spermatozoon is able to pass into the egg even if more sperms have reached the surface and the others are discarded because the cytoplasms in their respective entrance cones do not reach upto the heads. When the sperm is entering the egg the first polar body is extruded before the fertilization membrane is formed. After this the fertilization membrane is formed around the egg and the second polar body emerges at the same place as the first polar body. Finally, the first polar body remains above the membrane, and the second below it before the first cleavage. The first cleavage is total and equal

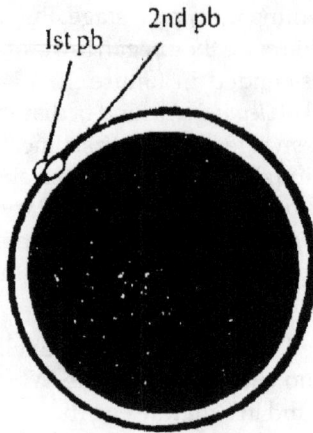

Figure 22: Egg of *P. japonicus*
Before Cleavage
First (1st pb) and Second (2nd)
Polar Bodies

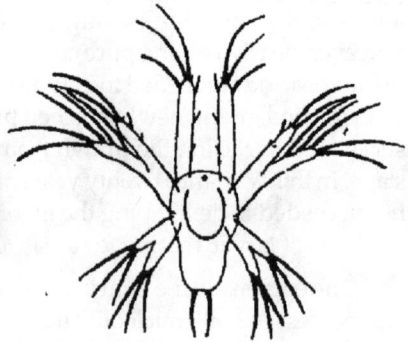

Figure 23: First Nauplius of
P. japonicus

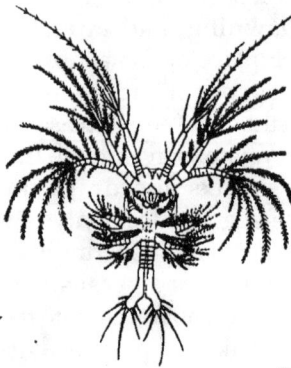

Figure 24: First
Zoea of
P. japonicus

Figure 25: First
Mysis of
P. japonicus

Figure 26: First
Post-Larva of
P. japonicus

and takes place 30 to 40 minutes after spawning. Subsequently further segmentation takes place every 12 to 15 minutes. Four cells are formed after the second division, and thereafter by further cleavage 64 to 128 cells are formed. An invagination is formed at the vegetative cells of the embryo. Soon an embryonic membrane surrounds the embryo after about 2 hours and it gradually becomes oval.

The rudiments of the antenna first appear as a swelling in the middle part of the embryo. The root of the mandible appears below that of the antenna followed by swelling of the antennule above it. When the three pairs of appendages have developed well, the embryo becomes a *nauplius*. A dark red ocellus appears on the anterior end of the body which is ventrally placed. The embryo begins to show signs of movement within the egg–membrane and in 13 to 14 hours after spawning the nauplius emerges from it.

The newly emerged nauplius larva has the three pairs of appendages; the anterior uniramous antennules, middle antennae and posterior mandibles, both biramous. It swims by means of the appendages, the antennae being the most active, the antennules comming next. They are attracted to light but avoid direct sun light. There are six nauplier stages, each stage formed after the moulting of the previous stage, their size ranging from 0.34 mm in the first to 0.51 mm in the sixth stage. After each moult body size increases and structural complexities take place. At rest, the nauplius keeps its dorsal side down, remaining in water in a perpendicular position with the three pairs of appendages slanting upwards. While hatching the colour of the embryo is dark yellow, and the colour becomes whitish and semi-transparent later by moulting. Appendages get some reddish-brown specks. Nauplii are not feeding stages, there being yolk in the body for nourishment. When the zoea stage is formed the yolk is completely absorbed.

The sixth nauplius moults in about 36 to 37 hours when the first zoea emerges. The body is elongated and there are changes in the appearance and swimming behaviour. The endopodites and exopodites of mandibles fall off, their function now being mastication. The first and the second antennae are the chief organs of locomotion and they are also aided in swimming by the first and second maxillipeds which have well developed. Zoea is photopositive but avoids very bright light. The carapace covers the anterior side of the body upto the eighth segment. The compound eyes make their appearance towards the end of this stage. The uncovered lower part has six segments, which belongs to thoracic region. Later five abdominal segments develop with distinct boundary lines. The telson is well developed with a semi-spherical forked end. The appendages are well developed with distinct articulations and plumose setae and spines.

There are three zoea stages formed by moulting, and after each moult more appendages and other structures are formed. Eyes protrude out in the second zoea stage. Mysis stage formed after the moult of third zoea. The mysis resembles the adult prawn having the cephalic and thoracic segments united to form cephalothorax which is covered by the carapace. Rosturm develops to a little more than half the length of the carapace.

The first and second antenna cease to be natatory and become olfactory in function. The five pairs of pereopods take on the function of swimming, with the assistance of three pairs of maxillipeds. They swim with their heads down and telson up, keeping the body in a slanting position. They do not show much preference to the light. The body colour is pale yellow, the colour of the mouth parts, thorax, abdominal segments and the tip of first antenna is reddish brown.

The size of the mysis varies from about 3.10 mm length in the first to 4.52 mm in the third stage. The abdomen is little shorter than the body. On the basal segment of the antennule, the rudiment of an otolith is developed. The rudiments of the green gland appear on the protopodite of antenna. On the first three pereopods of early mysis, rudements of arthrobranches appear and they become clearly distinguishable in the third mysis. In this stage a rudiment of arthrobranch appears on the fourth pereopods also. The five pairs of pleopods on abdomen are rudimentary and functionless. There are three mysis stages formed by moulting, and after each moult the body structures change. The post-larva emerges after the third mysis moults.

In the post larva the five pairs of pleopods become fully developed and functional as the chief organs of swimming, the uropods assist in balancing. The pereopods are used only for walking and grasping. At this stage the post larva goes down and creeps on the sand below. Day and night they creep and burrow into the sand at intervals. After 10 or 12 moults they begin to creep on the bottom

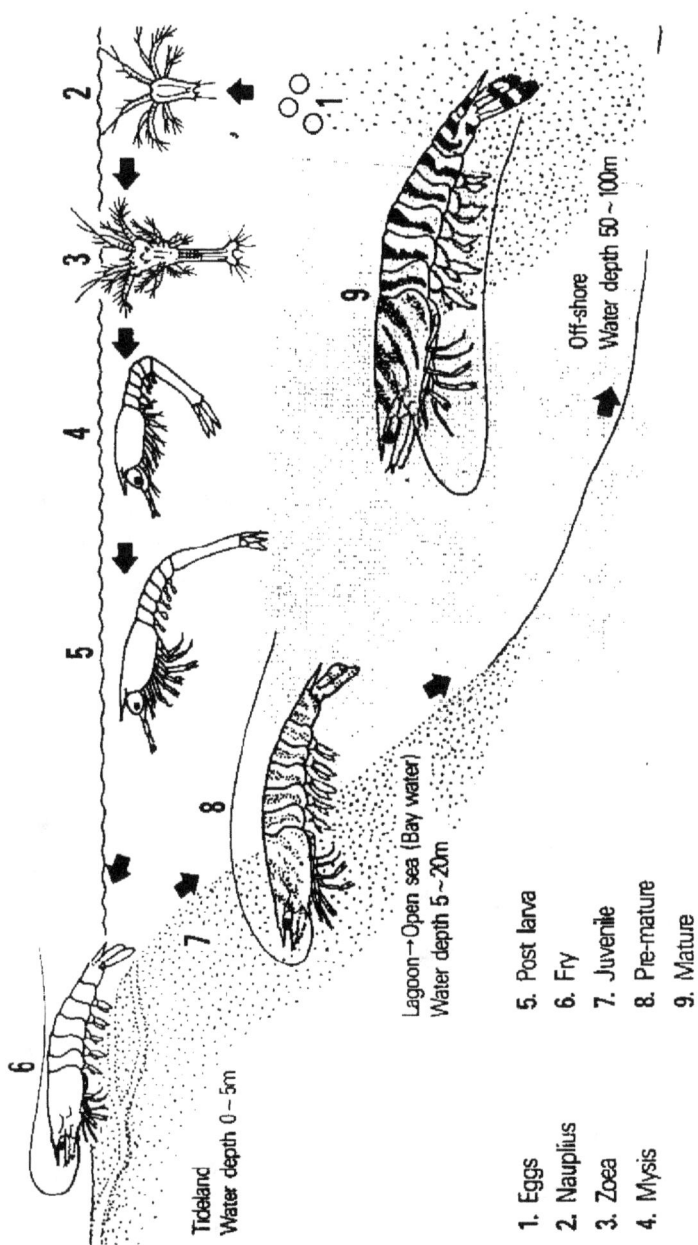

Tideland
Water depth 0 – 5m

Lagoon→Open sea (Bay water)
Water depth 5~20m

Off-shore
Water depth 50 ~ 100m

1. Eggs
2. Nauplius
3. Zoea
4. Mysis

5. Post larva
6. Fry
7. Juvenile
8. Pre-mature
9. Mature

Figure 27: Life Cycle of *Penaeus japonicus*

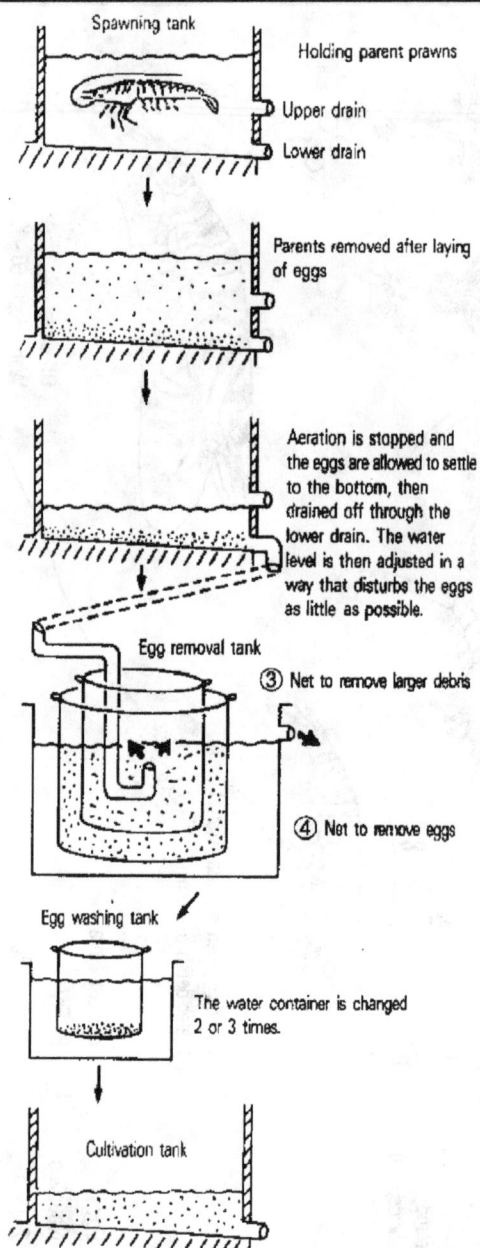

Spawning tank

Holding parent prawns

Upper drain

Lower drain

Parents removed after laying of eggs

Aeration is stopped and the eggs are allowed to settle to the bottom, then drained off through the lower drain. The water level is then adjusted in a way that disturbs the eggs as little as possible.

Egg removal tank

③ Net to remove larger debris

④ Net to remove eggs

Egg washing tank

The water container is changed 2 or 3 times.

Cultivation tank

Figure 28: Cleaning Operation of Fertilized Eggs

Figure 29: Parent Prawns (Spawners)

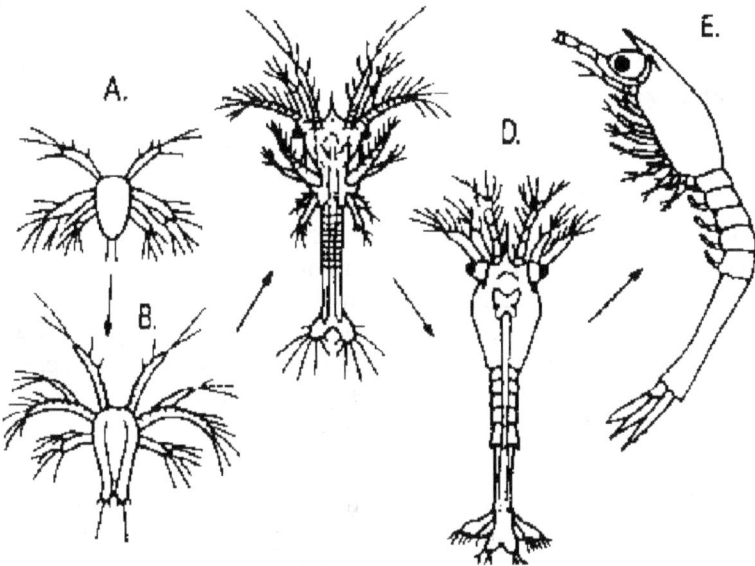

Figure 30: Kuruma Prawns Fry (*Penaeus japonicus*)
A: Early period nauplius; B: Middle period nauplius;
C: Middle period zoea; D: Early period mysis;
E: Middle period mysis

like adults. After every moult, they resemble more and more the adult form and after 20 to 22 moults the shape of the body and appendages resembles those of the adults. The colour changes in the post larva after 4 or 5 moults to greenish black, brownish black, brownish yellow on various parts of the body. The colour spots increase in number and when the prawn is 6 cm long gets its special adult colour.

The genus *Penaeus, Metapenaeus* and *Parapenaeopsis* form majority of commercially important penaeid prawns of India. All of them are basically marine. They occur throughout the Indian coasts at all depths, on the continental shelf and beyond. These prawns breed in the sea when they are of 6 to 8 months old at different depth zones depending on species.

Most of the species are continuous breeders with definite peak periods of breeding which vary from east to west coast. The eggs after spawning undergo a brief period of embryonic development and hatch out as nauplii. This larval stage passes through other stages, such as zoea and mysis and completes the larval metamorphosis in the sea. The mysis further transforms into the post larval stages and then to juvenile in the sea or in the adjoining estuaries and backwaters, which are used as nursery ground by some of the estuary dependent species. Around attainment of maturity the prawns return to the sea for breeding.

The number of eggs laid varies from 20000 to 10,00,000 depending on the species and size of the female. Depending on the temperature, the fertilized eggs hatch out to nauplii within 8 to 14 hours. There are six nauplier stages, three zoeal stages and three mysis stages before the larvae metamorphose to first post-larvae. Nauplier stages take 2 days, zoeal and mysis stages, each 3 days for development. Thus first batch of nauplii take 9–10 days to pass through different stages and metamorphose to first post-larvae. In case of species belonging to the genus *Metapenaeus* and *Parapenaeopsis* in addition to the usual three mysis stages, 2 to 3 substages are also noticed.

Penaeid eggs are demersal and are shed free in the water. Fertilization is external and is marked by the development of a perivitelline space. A narrow perivitelline space (15 micron) is the characteristic of the genus *Penaeus*. For *Metapenaeus dobsoni* and *P. stylifera*, the perivitelline space is 85 and 60 micron respectively.

In the eggs of *Metapenaeus* sp (except *M. dobsoni*) the perivitelline space is between 20-30 microns.

Larval stages of *Penaeus indicus*

Nauplius

Body pear shaped with three pairs of natatory appendages. No spines or processes are present on the antennae and mandibles for feeding purposes. The mouth is not formed and the larva is dependent on internal yolk for development. Pairs of caudal setae of equal length is extending straight posteriorly. Six well defined sub-stages of nauplius can be identified based on number of furcal setae. Non-plumose setae is present in first nauplius and the antennule bearing two long terminal and one long lateral setae. From second nauplius onwards the setae become plumose.

Zoea

A large carapace followed by a slender thorax and abdomen. Carapace does not cover the thorax completely. Uniramous antennules and biramous antennae with fully segmented exopods are present. Abdomen bifurcate posteriorly, each furca with at least seven setae. Three distinct substages of zoea can be identified.

In the first zoea, eyes are sessile. Spines are absent in rostrum and supraorbital region. Pereopods not present and the abdomen is unsegmented. In the second zoeal stage, eyes are stalked, rostral and supra orbital spines appear. The first five abdominal segments can be demarcated, but telson is not demarcated from the last abdominal segment. Uropod is absent. In the third zoeal stage uropod is present. Telson is seperated from last abdominal segment.

Mysis

Carapace covers the thorax. Third maxillipeds and the five pereiopods are functional with well developed exopods. First three pereiopods are with rudimentary chela. Pleopods if present are rudimentary. Three substages can be identified in the case of *Penaeus indicus*.

In first mysis, rostrum is long, extending beyond the eye and devoid of rostral spines. Supra-orbital spine is prominent. Hepatic spine is well developed. Carapace covers thoracic region completely and thoracic appendages are well developed. Posterolateral spines

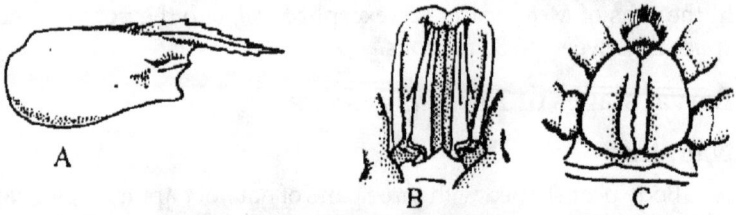

Figure 31: (A) Carapace; (B) Petasma and (C) Thelycum
of *Penaeus indicus*

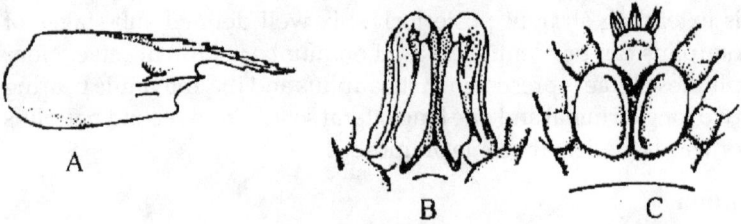

Figure 32: (A) Carapace; (B) Petasma and (C) Thelycum
of *Penaeus monodon*

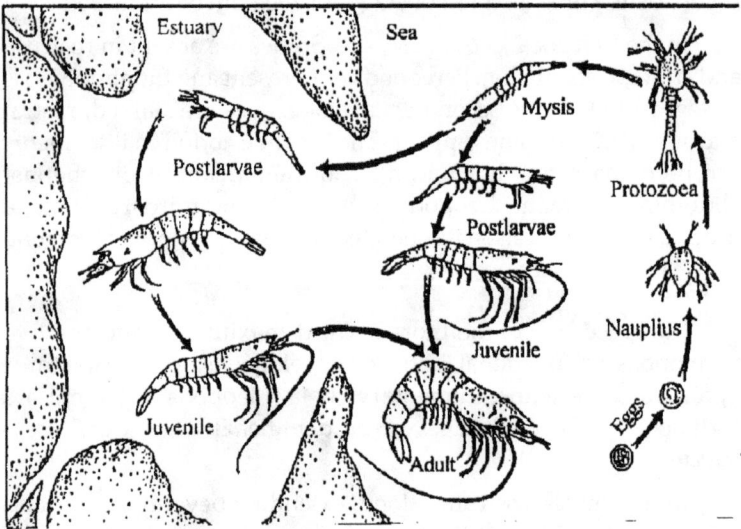

Figure 33: Life Cycle of Penaeid Prawn

Figure 34: Antarctic Krill, *Euphausia superba*

are present on fifth and sixth abdominal segments. A distolateral seta is present on the outer margin of exopod of antenna. Pleopod buds are not developed. In the second stage of mysis, a spine on the distolateral margin of exopod of antenna is present, though pleopod buds not developed. Two segmented pleopod buds without any setae developed in third mysis stage.

Unlike zoea, clear-cut morphological changes are not very clear in the substages of mysis. Last mysis moults some times one or two intermediate stages before metamorphosing to the first post-larva. Pleopods and setose are functional in first post larva. Exopods on pereiopods and second and third maxillipeds are lost and that of first maxilliped is reduced.

Parapenaeopsis hardwickii in Bombay coast breed throughout the year with high breeding and spawning activity during January–May. The recruitment period of the species to the Bombay coast are June to August and September–December when sufficient quantities of immature prawns join the coastal stock.

The oogenesis and ovarian maturation of *P. hardwickii* are orderly processes. The germinative zone is located in medio-ventral portion of the ovarian lobes. Oocyte maturation results in vesicle formation, eventual deposition of yolk and mature oocytes surrounded by follicle cells, proximate towards the peripheral wall of the ovary.

Depending upon the ovum size, gonad expansion plus the colour of ovary the degree of sexual maturity is categorized into 5 successive stages.

Immature (Stage–I)

The ovarian lobes are translucent or white in colour. Ova small of 30–50m and are transparent with clear nuclei.

Previtellogenic (Stage–II)

Ovary less translucent as compared to earlier stage. The oocyte diameter ranged between 169-190μ. Opaque yolk granules are formed in the cytoplasm and partly obscure the nucleus.

Vitellogenic I (Stage–III)

The ovary is light green and is visible through exoskeleton. The ovarian lobes are fully developed and their diameters are much larger than of the gut. The maturing ova are opaque due to the yolk accumulation. Oocyte diameter ranges from 235–260μ.

Vitellogenic–2 (Stage–IV)

The ovary is dark green or greenish brown. The ovarian lobes are filling the entire space inside the dorso-lateral surface of cephalothorax and abdomen. Oocyte diameter measured 290–325μ or even more.

Spawned-spent-recurring (Stage–V)

After extrusion of eggs, the ovary reverts almost immediately to the immature condition. Ovarian lobes flaccid and much convoluted. Colour varies from cream to light-yellow. Ova are small as in stage–I. This stage is distinguishable from that of stage I only by the size of prawn.

Most of the penaeid prawns of Indian coast breed in the inshore grounds. The larger species, such as, *Penaeus indicus* and *Metapenaeus monoceros* breed in slightly deeper waters, while the smaller species like *Metapenaeus dobsoni* and *Parapenaeopsis stylifera* do the same in shallow waters.

The vast majority of population belonging to *Metapenaeus dobsoni* to not grow beyond 60-65 mm in length in the backwaters. Those growing beyond 70mm are quite few and have seldom formed more than 2.5 per cent in any month. Even those belonging to the 66-70mm group make up only a small proportion (0.3-8.2 per cent)

The breeding period of *Metapenaeus dobsoni* along the Malabar coast is fairly long extending from September to March or April. In a course of a year and some months a good portion among them have grown to a length of about 80 mm. In the commercial catches two age groups, namely, 56-60 mm and 91-95 mm are represented in September or 101-105 mm and 116-120 mm are represented in November. The two modes at 101–105 mm and 116–120 mm in November represent the majority groups among males and females respectively of the same age. In December there is only a single mode placed at 106–110 mm. The group measuring 56–60 mm belongs to the first year class (0–year class). The other 116–120 mm represents the third year class.

Marine catches at Narakkal from May to October reveals that males are at 51-55 mm, 81-85 mm, and 101-105 mm classes. The frequencies of females also show the same three year-classes by the modes at 61-65 mm, 81-85 mm or 86-90 mm and 111-115 mm.

The data derived from rearing experiments and length frequency studies clearly show that prawn of this species (*M. dobsoni*) live for about three years and the males grow to 70 mm in the first year, 90-95 mm in the second year and about 110 mm in the third year, while female prawns grow to 75-80 mm, 100-105 mm and about 120 mm in the first, second and third year respectively.

The proportion of females is slightly higher (51.3–52.9 per cent) than that of males (47.1–48.7 per cent) in marine environment.

The breeding season of *M. dobsoni* is May to December within the sea and reaches its peak during August to December. The occurrence of fry in the canal and backwaters has actually due to their migration from the sea. Throughout the breeding season therefore, immense number of these fries (4-6 mm) would be continuously passing into the back waters from the sea.

The Indian marine prawns, *Penaeus indicus, Metapenaeus dobsoni, Parapenaeopsis stylifera* and *Penaeus monodon* have been induced to mature and spawn in captivity by using the technique of unilateral eyestalk ablation.

For removal of the eyestalk an electrocautery apparatus was used, as ablation with a pair of scissors led to profuse bleeding and heavy mortality. Cauterisation prevented bleeding and there was practically no mortality. No antibiotics were used to treat the ablated prawns. In the early experiments in which both the eyes were ablated, the development of the gonad was very rapid but mortality was very high and even fully mature females did not release the eggs. So unilateral eye stalk ablation was resorted to and this was proved very effective. The males were not ablated as they were fully mature even in the brackish water environment.

The ablated females were kept along with an equal number of males in a 12 feet diameter plastic lined pools in which seawater was recirculated through a sub-gravel filter by air-lifts. The depth of the water in the pools was 75-90 cm. The pools were accomodated in a tile-covered shed without side walls and no artificial illumination was provided. *M. dobsoni* and *P. stylifera* were kept in 6 feet diameter plastic pools provided with three air stones, the depth of sea-water being 50 cm. The animals were fed with fresh clam (*Sunneta scripta*) meat. Live mysids were also introduced into the pools whenever available. The waste products and uneaten food were removed every day by siphoning.

The temperature of the water of the pools varied from 24.5° C to 30.2° C and the salinity from 26.8 ppt to 38.6 ppt. The pH varied from 7.6 to 7.9.

The fully developed ovary of the mature *Penaeus indicus, M. dobsoni* and *P. stylifera* could be clearly seen through the thin cuticle without removing the animals from the pool. But in the case of *P. monodon* the ovary could not be seen easily through the thick pigmented cuticle. Hence they had to be caught every week and examined against a bright source of light to determine the degree of development of the ovary.

The females with fully developed gonads were removed from the pool with a dip net and placed individually for spawning in a 50 litre plastic basin or 3 feet diameter plastic pools containing filtered sea-water with good aeration. Spawning took place either in the same night or on the subsequent night. After spawning the females were removed and the number of eggs spawned and subsequently the number of hatched nauplii were estimated by counting aliquot samples.

M. dobsoni and *P. stylifera* attained maturity after eyestalk ablation even in a 6 feet diameter pool containing sea water which was ketp aerated with air stones. But *P. indicus* and *P. monodon* did not attain maturity under these conditions. Positive results were obtained with these two larger species only after the sub-gravel filter and air-lift circulation was fitted in the pools. It is clear that the quality of sea water is of prime importance in the maturation of these prawns. Recirculation of sea water through gravel filters, otherwise known as biological filters, is a well known method of converting the toxic ammonia excreted by the animals in the culture tanks to harmless nitrates by the nitrifying bacteria that grow on the gravel surfaces.

P. monodon appears to take a longer time to mature in captivity (sometimes upto 66 days). But *P. monodon* takes generally 24-25 days to mature after unilateral eye-stalk ablation. But *P. indicus* which resembles *P. merguiensis* in maturing 11-15 days after eye-stalk removal. The only viable spawning occured in the specimen which matured 10 days after eye stalk abalation.

None of the specimens of *P. indicus* below 30 mm carapace length (CL) responded to this treatment. The percentage of prawns which matured after eyestalk removal was relatively high among larger sizes. The minimum size of female which showed a positive response after eyestalk ablation was 52 mm CL. In *P. monodon* however, this tendency was less marked.

Generally the prawns attained maturity faster when the temperature of the pools was 28.5°C to 30.2°C and the salinity 30-33 ppt.

Viable nauplii could be obtained only during the period of December 1978 to May 1979, when the quality of sea water was good. During the rest of the period, which coincided with the south-west and north-east monsoon season, either the sea water was very turbid or it contain swarms of harmful dinoflagellates such as *Gymmodinium* sp and *Gonyaulax* sp.

Mating occurs at night, shortly after moulting while the cuticle is still soft and sperms, kept in a spermatophore is inserted inside the closed thelycum of female. Fertilization is external with female suddenly extruding sperm from the thelycum as eggs are laid in offshore water. Hatching occurs 12-15 hours after fertilization. The larvae, termed nauplii, zoea, mysis and early post larvae are free swimming. Nauplii feeds on reserved yolk stock and the rest at the larval stages are planktonic. The post larvae subsequently change their habit to feed on benthic detritus, polychaete worms and small crustaceans.

Due to better survival, large size and short culture period of *P. monodon,* wild seeds are generally captured and used in south Asia for extensive ponds, which require a minimum quality of seeds for stocking. Due to overfishing and outbreak of while spot syndrome virus, most grow-out farms now rely on hatchery-produced *P. monodon* seeds.

Healthy females (25-30 cm body length) and males (20-25 cm length) are captured from 60-80 m depth and 20 miles offshore sea in live condition. They are transferred to the maturation tank, which is normally covered and kept in dark room. After the completion of mating, which is easily known by the presence of spermatophore in the thelycum and hardening of the shell, the eyestalk of females is unilaterally ablated for endocrine stimulation.

It was generally observed that the post larval incursion was maximum during the highest tides (Spring tides) of full moon and new moon quarters. Most post larvae were entering the lake at night than during the day. The post larvae of *P. indicus* entered the lake in unusually large numbers in the day time on the full moon day (12186/hour) while the incursion at night on the same day was relatively low (5284/hour). In all the other months distinct diel periodicity was evident. A negligible number moved out at ebb tide.

Postlarval incursion is relatively rich at night, and was maximum during the third, fourth and fifth hours after the setting of the tide in case of all the three species. In case of *P. indicus,* however, the incursion of post larvae was considerable even during the second hour. It was also observed that the strength of the current was maximum during the period of the peak post larval recruitment.

The juveniles of *Penaeus indicus* attained an average size of 92.7 mm in two months, and 100.44 mm in three months. It has been observed that captive post larvae feed on algal matter but grow faster if animal supplement is given. The post larval growth of *P. indicus* is maximum for about three months and decreases gradually thereafter. The rate of growth is very low after the fifth month by which time the prawn attains an average size of 114.87 mm.

The post larvae of *P. indicus* showed abundance during two periods, January to April and June to September. In the Cochin backwaters also two peaks of incursion were noted, one during February to April and the other in November to December. The former period is closely similar to the first peak period in the Pulicat Lake and also in the Chilka Lake. The period of occurrence of the second peak is different in Cochin backwaters, as the period coincides with the north-east monsoon in the Pulicat Lake.

The availability of prawn fry at the mouth region alone does not appear to be peculiar to the Pulicat Lake. In the Chilka Lake also the availability and abundance are often confined to the lower most regions of the outer channel. It was also noticed that post larvae of prawns are more abundant near the mouths of tidal lakes and rivers than off shore or well up in tidal streams.

The post larval studies are very useful to predict the nature of the future prawn catch. These predictions are however, limited to indicate only good or bad seasons. The number of post larvae recruited during the incursion period (January to April and June to September in Pulicat lake) are considered as indicators of future prawn abundance. The only difference between the Chilka Lake and Pulicat Lake appear to be poor post larval incursions during the period June to September, as a result of which the secondary prawn production peak (October to December) is not significant in the Chilka Lake.

Of the penaeid prawns, *Penaeus monodon* and *Penaeus indicus* are the most important from the culture point of view, due to their availability throughout the brackish water areas of the country and their attaining big sizes of about 300 and 200 mm total length respectively.

The post larvae of three *Penaeus* species, namely, *Penaeus monodon, Penaeus indicus* and *Penaeus semisulcatus* can be identified on the basis of chromatophore distribution and other characters.

In *Penaeus indicus* post larvae five to seven reddish brown (or yellowish) chromatophores are present on the ventral side of the sixth abdominal somite. One reddish brown chromatophore is visible at the anterior end of the sixth abdominal segment laterally.

In *Penaeus semisulcatus* post larvae more than eight (8-11) reddish brown (sometimes bluish) chromatophores are present on the ventral side of the sixth abdominal somite. One reddish brown chromatophore at the anterior end of the sixth abdominal segment is found laterally. One or two reddish brown chromatophores are also present on the dorsal side of each abdominal segment.

Forteen to nineteen reddish brown (sometimes bluish) chromatophores are present on the ventral side of the sixth abdominal segment of *Penaeus monodon* post larva. No lateral chromatophore is found on the sixth abdominal segment anteriorly. The ventral chromatophores on the sixth abdominal segment appear as a bluish or reddish brown streak in expanded condition.

The presence of both ventral and dorsal teeth on rostrum, visible to the naked eye from the later juvenile stages, distinguishes the genus *Penaeus* from the other penaeid genera, *Metapenaeus* and *Parapenaeopsis*.

Juveniles of *Penaeus monodon* can be readily distinguished from those of *Penaeus indicus* as well as the other commercially important penaeids by the presence of a red streak that passes all along the ventral side of the body (upto 14 mm in total length) from the rostrum to the tip of the telson. The colouration is seen in the antennule also. In slightly bigger individuals, (15 to 20 mm total length), the colouration of the streak progressively changes into pink and then green. By about 20 mm length the pink and green colouration spreads to the entire body with the streak prominent on the ventral side. In longer specimens the colouration of the entire body becomes greenish.

Penaeus indicus is more or less transparent in the early juvenile stages, but from about 25 mm onwards, the tip of the rostrum becomes tinged slightly red or pink. The eyes of *P. indicus* juveniles are generally bigger in size with slightly longer eye-stalks, as compared with those of *P. monodon*. This is an additional confirmatory character for seperation of live juveniles (10-20 mm size) of the two prawn species in the field.

In *P. indicus* juveniles of 10 mm length onwards four or more ventral teeth are normally noticeable, while in *P. monodon*, they may be only two in number till 20 mm stage, when the third teeth is formed. The rostrum of *P. indicus* juveniles were found to be comparatively longer than that of *P. monodon*.

The Pulicat Lake is a marine dominated brackish water system on the east coast of India and is rich in prawn fisheries comprising about 40-50 per cent in the total landings from the lake. The total prawn landings from the lake are about 476 tonnes. Although several varieties of prawns are recorded from the lake, a single species,

Penaeus indicus contributes the bulk of the prawn catch (300-400 tonnes per year) comprising about 60 per cent of the total prawn landings from the lake. The other two species, *Penaeus monodon* and *P. semisulcatus*, do not contribute much to the commercial catch, but the size of the prawns landed makes them desirable for the prawn freezing plants. All the three species grow fast and reach marketable size in 4 to 6 months.

In the mouth region of Pulicat Lake, connecting the adjacent sea, the post larvae of *Penaeus indicus, P. monodon* and *P. semisulcatus* were recorded almost throughout the year, those of *P. indicus* being the dominant group are very plentiful during two times of the year, January to April and June to September. The post-larvae of *P. monodon* are usually rich during January to April and August to November. The post larvae of *P. semisulcatus* are plentiful during March to June and September to October.

Prawn's Diet

The prawns eat all types of plants and animals, living or dead that comes in their way. Since they inhabit in the benthic region they feed on all living animals and vegetable matter available at the bottom including dead organic matters. Food is grabbed by the chelae of the thoracic legs and taken to the mouth where the maxillipeds help in further cutting and driving it towards the mandibles which macerate the food into finer particles fit for swallowing. The food of young penaeids consists of algae, minute organisms and organic detritus. While examining the gut contents of *Penaeus indicus* from the inshore waters of different size groups (11–14 cm; 14–17 cm and 17–20 cm) the chief contents of the food were found to be digested matter and detritus, vegetable matter, crustacean and non-crustacean animal matters. Vegetable matter constituted fragments of different kinds of algae, like, *Trichodesmium* and also cuttings of *Pleurosigma* and *Rhizosolenia* are found. Crustacean remains include small parts of copepods, ostracods, amphipods and tiny decapods. Among non-crustacean alimal matters molluscan shell particles, and broken pieces of some types of larvae is encountered. Parts of hydroids, and echinoderm larvae and sometimes living trematodes are present. Sand particles enter through the food items. Prawns have a wide adaptability in their feeding habits, eating whatever suitable material happens to fall in their way.

Prawns kept under observation was found to consume algae, planktonic organisms, muscle pieces and polychaetes preferring smaller ones, which could be held in the chelae of legs. For bigger pieces more than one chelate legs are used. Bigger prawns normally do not attack smaller ones except when they become weak or dead.

The diet of *Metapenaeus affinis* of 106 to 165 mm, predominantly are made of animals living in the bottom, such as, nematodes, foraminiferans and molluscs. Gastropods and bivalves form about 11 to 99 per cent of food. Older prawns prefer molluscan diet. Nematodes, copepods and amphipods form lesser in proportion to Molluscs. Diatoms and algal filaments were found only in few prawns. Debris and sand particles were observed along with other items of food. Post larvae feed better on mixed diet of algae and nauplii and a pure animal diet is better than pure plant diet. The mud contains bacterial colonies and filamentous algae in which protozoans, copepods and nematodes and other fauna are abundant. In a densely populated shrimp grounds large food masses may be inadequate to support a healthy population and this deficiency is likely to be made up by feeding on the epifauna and epiflora of the mud surface. During the process of browsing, mud can also pass into the stomach. The tufts of setae of the chelate legs can detect the presence of food particles on the substratum.

Small crustaceans form the major food item of juvenile prawns in Cochin backwaters. With the increase in size selective feeding is more evident among marine prawns. *Penaeus indicus* feed mainly on large crustaceans, where as *Metapenaeus monoceros* on smaller crustaceans. *M. affinis* was found to feed on vegetable matter and *M. dobsoni* to be a detritus feeder.

The food of penaeid prawns in general, consisted of varying amounts of organic matter mixed with sand and mud in Cochin area. Unidentifiable 'debris' was on an average was more in *P. indicus* and less in *M. monoceros*. In all species greater amount of debris was found in younger than in adult specimens. *M. monoceros* feed on more vegetable matter than the other species. In prawns caught from estuaries the food contents comprised of polychaetes, amphipods, bivalves, gastropods, foraminiferans, nematodes, sand and mud, exhibiting a marked benthic feeding. In specimens obtained from paddy fields, stomach content shows large amount of plant matter. Food contents also show variations in different seasons in different species. In *P. indicus* algal contents were higher in the monsoon season and molluscan remains and crustaceans in post-monsoon periods. In *M. dobsoni* crustaceans are more common in monsoon seasons, whereas, in pre-monsoon and monsoon months plant matter in on increase, becoming lesser in post-monsoon months. Crustacean remains are more predominant in the diet of *P. monodon* during pre-monsoon and monsoon months. Selection of food by the various species is evident inhabiting in the same environment.

Prawns have food preference for crustaceans, or molluscs or algae and in case of unavailability of their usual food, they go for other items. The food content of different species or of the same species may vary according to environment and seasons. When feeding on bacteria or algae, lot of mud and sand also enter their stomach. Inspite of all those, marine prawns do have the selective instinct for food they consume.

Food and Feeding Habits

Considerable differences have been noticed in the food preferences of various larval stages, juvenile and adult prawns. The larvae in the nauplius stage do not feed at all, as they have food reserve in the form of yolk. But protozoea larvae feed voraciously on phytoplankton such as, *Tetraselmis* spp, *Synecocystes* spp, *Thalassiosira* spp, *Skeletonema costatum* etc. It has been observed that Indian species namely, *Metapenaeus affinis*, *M. dobsoni* and *Parapenaeopsis stylifera* thrive well on cultures of the diatom, *Thalassiosira* spp registering very high survival rates. In the mysis stages they feed well on brine shrimp nauplius and yeast granules.

Adult prawns feed on a variety of animals and plant material available in the area where they live. They feed on crustaceans, polychaetes, molluscs, radiolarians, forameniferans, pisces, diatoms, algae etc. along with considerable quantities of organic detritus from the bottom of the sea or backwater. It has been observed that *P. semisulcatus* are actively feeding in the night and that the quantity of food items varied considerably depending on the abundance of these items during the seasons. It is evident that they are in general not selective feeders.

There has been individual variations among the species. *P. indicus* feed on plant material in the younger stages while the older ones prefer a predominently crustacean diet. Algal filaments also form part of the food of this species. *P. monodon* feed on molluscs crustaceans, polychaetes and fish remains. *P. semisulcatus* consume large quantities of animal matter, namely, polychaetes, crustaceans, molluscs, radiolarians, forameniferans, fishes, as well as diatoms and algal filaments. *M. dobsoni* feed mainly on crustaceans, algal filaments, seaweeds and diatoms, such as, *Fragilaria, Coscinodiscus, Pleurosigma, Biddulphia* etc. *M. monceros* eat crustaceans including copepods, ostracods, amphipods and decapod larvae in addition to plant materials like sea weeds and algae.

The oesophagus of prawn is a short tube continuing vertically into anterior chamber of the proventriculus (stomach) which extends backwards as the gastric mill. The groves on the floor of the anterior chamber serves for the passage of the digestive secretions from the digestive gland which get mixed up with the food ingested. The posterior chamber of proventriculus is partly embedded in the digestive gland and is divided into a dorsal channel which leads directly into a long simple mid-gut and a ventral filter press which allows only fine particles to pass into the digestive gland.

The digestive gland is compact in shape, forming about 3-4 per cent of the total weight of prawn. There are two openings from the ventral filter press directly into the simple tubules which constitute the gland. The tubule walls are formed of a single layer of epithelial cells distinguished as the "secretary cells" and "storage cells", the latter containing polysaccharides. Proteinases which are enzymes in the digestive juice are reported from *P. orientalis*. The rapid postmortem autolysis of the digestive gland found to occur in prawns due to these enzymes. Amylase equivalent to 40 units per gram is also found in the digestive gland. It is probable that lipases may also be present as the fats are emulsified by the digestive glands.

By the time the food reaches the digestive gland, digestion is well under way and is completed in the proximal half of the digestive gland tubules. The larger indigestible particles are passed backwards through the dorsal part of the posterior proventriculus into the midgut from where they are expelled as faecal matter through the rectum when the midgut is full. The anterior and posterior diverticula of the midgut is responsible for the osmotic and ionic regulations and have no digestive function.

On feeding enough soft food to a starved prawn, the anterior chamber of proventriculus gets filled up to capacity with in 2 minutes.

Defecation in starved animals start an hour after feeding reaching peak in 4-6 hours and ceases by 8 hours. The assimilation can be very rapid especially when the food is in very fine form. The rates of ingestion and digestion being more or less equal, the relatively small size of the stomach is not a major disadvantage.

Prawn Nutrition

Nutritional aspects of penaeid prawns attracted considerable attention when there was a big shift from capture to culture technology employed in prawn cultured practices especially in the Asian countries for high demand in the international market. It was all the more essential when along with the improvements of traditional culture practices, there has been a parallel development of semi-intensive, intensive and super-intensive culture technologies.

Production of prawns in extensive type of culture systems mainly depends upon natural food. Natural food production could be augmented by judicious application of certain organic and inorganic fertilizers at optimum doses. The overall productivity in the extensive culture systems depends on the natural biogenic potential, prevailing environment conditions as well as on the kinds and levels of fertilizers used. In general, production rates exceeding 500 kg/ha/crop of 4 to 5 months are rarely achieved without an exogenous supply of feed (artificial feed) in prawn ponds.

Nutrionally balanced complete feeds are essential in semi-intensive and intensive prawn grow-out systems. The provision of adequate levels of nutritionally essential chemical entities which play a significant role in the prawn either as tissue components or their precursors or as physiologically essential components or as fuels is the most important consideration in compounded feed design. A balanced feed formula should ensure a source of energy plus a blend of essential amino-acids, essential fatty-acids, a source of sterol (cholesterol) phospholipids (phosphatidylcholine + phosphiatidyl ethanolamine), vitamins and minerals to sustain life and promote maximum growth at optimum intake.

All the organic compounds in the food release heat upon combustion, but fats, proteins and soluble carbohydrates are the main energy yielding nutrients for prawn. All of these components play some part in the structure of the animal, but the need for energy can preclude their incorporation into tissues and lead to their catabolism. If the diet provides less energy to the animal than is needed to sustain the life processes and to support its physical activities, tissue components will be catabolised in addition to the food. If the energy/protein ratio is too low, protein will be used to satisfy energy requirements first. However, optimum energy/protein ratio in the diet for prawn species is yet to be established. In complete prawn diets metabolizable energy greater than 3200 k cal/kg has been suggested.

Protein is the first nutrient parameter in prawn's diet. Quantitative protein requirements have been established for a number of prawn species. Prawns require ten essential amino acids; arginine, histidine, lysine, leucine, isoleucine, methionine, phenylalanine, threonine, tryptophan and valine in the diet as they could not be biosynthesised *de novo* in adequate levels. Besides, tyrosine is biosynthesised from phenylatanine by a irreversible pathway. So tyrosine may become essential when phenylalanine levels are inadequate.

In general marine prawns require relatively high levels of protein of marine origin as compared to freshwater forms. Fish meal, calm meal, mussel meal, crab meal, prawn meal, squid meal either individually or in combination have been found to be superior to other animal and plant protein sources. Since, marine animal proteins being more expensive, prawn nutritionists are directing their research efforts to partly replace with plant and animal wastes and by-products. Among these soybean meal has been found to be an effective supplement. Slaughter-house waste, silk worm pupae, meat meal, meat and bone meal are also being incorporated in some diets. Groundnut oil cake, sesame seed cake and coconut oil cake are also used along with animal protein diets. With the exception of soybean, all other plant sources are inferior to animal protein sources in terms of their essential amino acid profile. While compounding diets, however, it is possible to select and utilize a variety of natural protein sources to achieve a more balanced amino acid profile. At present, amino acids balance in prawn feeds is achieved based on the essential amino acid profile of whole prawns or prawn tail muscle or the amino acid profile of the best natural food protein source, squid protein, calm protein etc.

Ammonia is the principal end product of protein metabolism in prawns and much of the ammonia is transferred across the gills is in the un-ionized state. Amino acids imbalance in diets could lead to increased ammonia production. Elevations of environmental ammonia above normal levels induce stress in prawns to pathogenic infections.

Lipids and Fatty Acids

Lipids are important nutrients in the diet of prawns as a source of energy. Essential fatty acids (EFA), sterol, phospholipids acts as carriers of fat soluble vitamins. In complete feed for prawns about 6 to 10 percent lipid containing a blend of essential fatty acids and phospholipids is essential.

Prawns have a requirement for polyunsaturated fatty acids such as, eicosapentaenoic (20:5n-3), docosahexaenoic (22:6n-3), linolenic (18:3n-3) and lenoleic (18:2n-6) acids with greater proportion of the first two fatty acids. Salinity and temperature of the environment are two factors that are known to affect the requirement of these fatty acids. Besides, the n-3/n-6 fatty acids ratio is also considered very important, with the n-3 fatty acids playing a dominant role. The major functions of EFA are related to their role as components of phospholipids and as precursors of prostaglandins. The EFA are found in highest concentration in phospholipids and as such are important in maintaining the flexibility and permeability of biomembranes, in lipid transport and in activation of certain enzymes. As precursor of prostaglandins they are probably involved in many diverse physiological and metabolic functions. Lipids of marine animal origin are found to contain high levels of 20:5n-3 and 22:6n-3 and are found to be superior lipid sources for prawns.

Prawns also require a source of phospholipids rich in phosphatidylcholine with some levels of phosphatidylethanolamine and phosphatidylinositol. Soylecithin at levels ranging from 1 to 2 percent has been found to promote growth in prawns. Prawns also require a sterol source in the diet and cholesterol has been found to have maximum efficacy. Cholesterol is important for synthesis of various steroids, vitamin D, brain and moulting hormones and also in hypodermis formation. A level of 0.25 to 0.5 percent is adequate in feeds. Marine invertebrate meals and oils, squid, shrimp, calm, crab, mussels, are excellent sources of cholesterol.

Carbohydrates

Carbohydrates are the least expensive form of dietary energy and may serve as precursor for various metabolic intermediates, such as non-essential amino acids, nucleic acids and chitin. Gelatinised starch helps bind the feed. Among carbohydrates, starch, dextrin, maltose and glycogen are effectively utilized by prawns. Gelatinised starch upto a level of 20 per cent in the diet is better utilized, though starch levels ranging from 35-40 per cent do not adversely affect the performance in certain species. Carbohydrate requirement of prawns could be satisfied by providing gelatinised wheat flour, maize flour, millet flour, cassava flour etc.

Chitin

In prawns chitin is required for the formation of exoskeleton as well as the periotrophic membrane in which the faecal pellets are enveloped. Though chitin is synthesised by the animal, inclusion of chitin (0.5 per cent) or its precursor glucosamine (0.8 per cent) has been shown to promote growth and improve feed efficiency. Prawns moult at frequent intervals and are known to feed on the moult. Thus a greater proportion of chitin is recycled. Exoskeleton from prawns, crabs and mantis shrimp can be used in the diet as sources of chitin.

Table 5: Recommended Nutrient Levels in Feeds for Penaeid Prawns

Nutrient	Size of Prawn (gm)			
	0–0.5	0.5–3.0	3–15	15–40
Protein (per cent)	45	40	38	36
Lipid (per cent)	7.5	6.7	6.3	6.0
Essential Amino Acids (per cent of feed)				
Arginine	2.61	2.32	2.2	2.09
Histidine	0.95	0.84	0.80	0.76
Isoleucine	1.58	1.40	1.33	1.26
Leucine	2.43	2.16	2.05	1.94
Lysine	2.39	2.12	2.01	1.91
Methionine	1.08	0.96	0.91	0.86
Methionine + cystine	1.62	1.44	1.37	1.30
Phenylalanine	1.80	1.60	1.52	1.44

Minerals

Minerals are important as constituents of the exoskeleton, structural components of tissues, in the regulation of osmotic pressure, nerve impulse transmission and in muscle contraction. Many of the minerals serve as components of enzymes, vitamins, hormones, pigments and as co-factors in metabolism or as enzyme activators.

Prawns moult at frequent intervals and the exoskeleton require a continuous supply of calcium and phosphorus apart from chitin. The dietary requirement for minerals is largely dependent on the mineral concentration of the aquatic environment in which the prawns grow. Since mineral are directly absorbed from the water via gills and body surfaces, it is generally believed that the requirement for many of the minerals could be fulfilled by absorption from the environment. However, phosphorus is considered to be essential in the diet as the natural waters are quite deficient in the element.

The calcium phosphorus ratio in feed is very important for the effective utilization. The recommended ratios range 1 : 1 or 1.5 : 1. The functions of dietary essential minerals and recommended levels in prawn's diet are given in Table 6.

Since the digestive system of prawns is not very acidic, mineral supplements that are water soluble are most available for prawns. While adding mineral supplements to feeds, the level and bioavailability of minerals from the complex natural ingredients should be considered so as to avoid any excesses. Ingredients, such as, fish meal, prawn meal, crab meal, squid meal and yeast are rich sources of most of the mineral elements.

Vitamins

Both water-soluble and fat-soluble vitamins are essential for prawns. At low stocking densities natural foods may be adequate enough to provide some or of the essential vitamins. But in high stocking densities, where natural foods are limited to support the entire population dietary vitamins are important. Vitamin requirements are affected by prawn size, age, growth rate, environmental conditions and the nutrient composition of diet. Loss of water soluble vitamins during processing and storage of

Table 6

Mineral Elements	Important Functions	Recommended Levels	Sources in Mineral Premix
Calcium (Ca)	Component of skeletal tissues, blood clotting, activation of enzymes, muscle contraction, cell permeability	2.8%, Ca-P ratio 1:1–1.6 : 1	Ca lactate Ca chloride Ca phosphate
Phosphorus (P)	Component of skeletal tissue, phospholipids, nucleic acids, phosphoproteins, high energy compound (ATP), many metabolic intermediates and co-enzymes.	Available P 0.9% Total P 1.8%	Na phosphate Mono calcium phosphate Di-calcium phosphate
Magnesium (Mg)	Component of exoskeleton, many enzymes associated with protein, lipid and carbohydrate metabolism, muscle and nerve functions and osmoregulation	0.2%	Magnesium carbonate
Sodium (Na) Potassium (K)	Found in body fluids and soft tissues, role in osmoregulation, acid-base balance and water metabolism	Na 0.6% K 0.9%	Na choloride K chloride K carbonate
Sulphur (S)	Component of sulphur amino acids methionine and cystine, glutathione, taurine, heparine, chondeitin sulphate		
Iron (Fe)	Utilized in enzymes–cytochromes, catalases, proxidases, oxidases, dehydrogenases	300 ppm	Ferrous gluconate Ferric sulphate
Copper (Cu)	Required in oxidation-reduction enzyme systems, component of haemocyanin	25 ppm	Cupric sulphate Cupric chloride

Contd...

Table 6—Contd...

Mineral Elements	Important Functions	Recommended Levels	Sources in Mineral Premix
Zinc (Zn)	Component of more than 80 metalloenzymes and a cofactor in enzyme systems	110 ppm	Zinc sulphate
Manganese (Mn)	Cofactor for number of enzymes including phosphate transferases and dehydrogenases, alkaline phosphatase, arginase and hexokinase	20 ppm	Mn sulphate
Selenium (Se)	Component of the enzyme glutathione peroxidase	1 ppm	Na selenite
Cobalt (Co)	Essential component of vitamin B_{12}	10 ppm	Co chloride Co-sulphate

ingredients and finished feeds and leaching from feeds when introduced into water are factors that affect the availability of vitamins to prawns.

Additives

Artificial feeds for prawns should have adequate levels of a binder or a mixture of binders to render the required water stability. Considering the binding capacity and cost, gelatin, collagen, alginates, and agar is used in levels ranging from 2 to 5 per cent; gum acacia, guar gum, carboxy methyl cellulose is included at a level of 1–2 per cent whereas, wheat gluten, alpha starch, wheat flour etc could be used in relatively high levels (30 per cent).

Antioxidants are essential to protect the fatty acids, and other oxidizable components in the feeds. They can be added to lipids or the vitamin premix during manufacture of the feed. Commonly used antioxidants are butylated hydroxy toluene (BHT) 0.2 per cent, butylated hydroxy anisole (BHA) 0.2 per cent and ethoxyquin 0.015 per cent. Natural antioxidents which go as components of feed include ascorbic acid and vitamin E.

Carotenoids, the substances which render colour to prawns are not bio-synthesised. β-carotene and its oxidative derivatives, crypoxanthin, zeaxanthin, canthaxanthin and astaxanthin which are commonly found in a number of invertebrates are the preferred carotenoids. Astaxanthin at a level of 50 ppm has been shown to improve colouration in prawns.

Free amino acids, small peptides and certain nucleotides present in fish solubles, fish meal, shrimp meal, squid meal, calm meal, polychaetes etc are known to induce feeding behaviour in prawns (Table 7).

Anabolic Agents

Two proteolytic enzymes, bromelain and papain at a level of 0.1 per cent and 0.2 per cent have been shown to promote growth in prawns. Olaquindox, a swine growth promoter, is believed to improve growth and feed efficiency at a level of 20 gram/tonne of feed, but residual effect is suspected. Bile salts included at 400 gram/tonne of feed has also been a growth promoter. Substances like zeolite (1-2 per cent), thyroproteins and phytosterols (0.1 per cent) are also considered to improve growth performance.

Table 7

Vitamin	Important Functions	Recommended Levels in Diet mg/kg	Sources for Premix
Water-soluble			
Thiamine (B$_1$)	Coenzyme for oxidative decarboxylation of pyruvic acid	150 mg	Thiamine mononitrate Thiamine hydrochloride
Riboflavin (B$_2$)	Coenzymes for flavin mono nucleotide (FMN) and flavin adenine nucleotide (FAD), involved with pyridoxine in the conversion of tryptophan to nicotinic acid	100 mg	Riboflavin
Pyridoxine (B$_6$)	Required for many enzymatic reactions in which amino acids are metabolised	50 mg	Pyridoxine hydrochloride
Pantothenic acid	Part of coenzyme A to transfer acyl groups in enzymatic reactions, role in fatty acid oxidation and synthesis, synthesis of steroids, cholesterol and phospholipids, pyruvate oxidation etc.	100 mg	Calcium di-panto-thenate (92 per cent activity) Calcium di-pantothenate (46 per cent activity)
Niacin	Component of coenzymes NAD and NADP required for synthesis of high energy phosphate bonds and serve as hydrogen acceptors	300 mg	Nicotinic acid or
Biotin	Intermediate carrier of CO_2 in several specific carboxylation and decarboxylation reactions, required in the biosynthesis of long-chain fatty acids, purine and the metabolism of odd-chain carbon fatty acids.	1 mg	D–Biotin

Contd...

Table 7–Contd...

Vitamin	Important Functions	Recommended Levels in Diet mg/kg	Sources for Premix
Folic acid	Involved in one carbon transfer mechanisms as in the metabolism of amino acids, biosynthesis of purines and pyrimidines	20 mg	Folic-acid
Inositol	Component of the inosital phosphoglycerides and inositol phospholipids	300 mg	Meso-inositol
Choline	Source of methyl groups, involved in a number of trans-methylation reactions, essential component of phospholipids and acetylcholine which has functions in lipid transport, cell structure and transmission of nerve impulses.	400 mg	Choline chloride
Ascorbic acid	Protector of enzymes and hormones from oxidation and inhibition, formation of collagen, RNA synthesis	1200 mg 250 mg	Coated ascorbic acid Ascorbic acid derivatives
Fat soluble			
Vivamin A	Involved in calcium transport, reproduction, cellular and sub-cellular membrane integrity	15000 I.U.	Vitamin A acetate or palmitate
Vitamin D	Involved in calcium and phosphorus metabolism and alkaline phosphatase activity	7500 I.U.	D_3 cholecalciferol
Vitamin E	Anti-oxidant, protects lipids of bio-membranes from oxidation	400 mg	di-tocopherol acetate
Vitamin K	Normal blood coagulation, energy metabolism, ameliorates aflatoxin toxicity	20 mg	Menadione sodium bisulphite

Added to improve shelf-life and to control deterioration due to fungal attack, sodium and calcium propionates at levels ranging from 0.1-0.25 per cent are recommended, although several other preservatives can also be incorporated.

In high-tech aquafeeds certain broad spectrum medicines, such as, oxolinic acid, oxytetracycline, terramycin etc are included to prevent out break of pathogenic infections. However, the inclusion of antibiotics in prawn feeds is not advised due to (i) the use of antibiotics may reduce bacterial populations in the aquatic environment that significantly contribute nutrients to shrimp and (ii) if fed routinely, resistant strains of pathogens could develop.

Growth and Maturity

The male and female *P. indicus* attain lengths of 128 and 143 mm at the end of the first year, 163 and 173 mm at the end of the second year respectively in the open sea. Prawns growing beyond 195 mm are rare and presumed to be 3 years old. Growth rate is about 15 and 20 mm per month in male and female respectively towards the end of the first year. Growth of juveniles of 30 to 100 mm is much faster when they are in the estuaries and backwaters. Greatest recorded size is 230 mm.

Life span of *M. dobsoni* is 2 to 3 years and greatest size attained is 128 mm for females and 118 mm for males. Growth rate is much faster in the backwaters than in the sea, where they grow to 60 to 80 mm in 7 or 8 months. In the open sea the, growth rates are 20 and 25 mm per month in male and female respectively during the first 6 months. They attain lengths of 70 and 75 to 80 mm, 90 to 95 mm and 100 to 105 mm and 110 and 120 mm in male and female respectively at the end of first, second and third year of life.

The life span of *M. monoceros* is 3 years and maximum size attained is 180 mm. They grow to 100 to 110, 131 to 135 and 156 to 160 mm at the end of first, second and third year respectively.

M. affinis live for 2 to 3 years only and their greatest size is 165 to 170 mm. In the trawl fishery the growth rate is 20 and 25 mm per month in male and female respectively. Length of 105 and 135 mm, 115 and 155 mm and 155 and 175 mm in male and female respectively are attained at the end of first, second and third year.

The life span of *M. brevicornis* is 3 years and maximum size attained is 127 mm. Lengths of 46 and 47 mm and 81 and 89 mm in male and female are attained respectively at the end of first and second year. After the postlarval phase it grows at the rate of 3 mm per month. Growth rate is faster in summer, less in rainy season and lowest in winter.

The life of a prawn involves periodic shedding of the old confining exoskeleton and subsequent enlargement of the newly disclosed integument. This process is called "moulting". It is a highly complex and profound physiological event in the life of a crustacean. Visible increment in size takes place at this time. The periodicity of moulting varies according to different conditions like light, temperature, food supply etc. Frequency of moulting depends on the age of the individual and the amount and quantity of food taken. Young specimens moult more frequently than old ones. Starvation usually inhibits moulting, where as feeding promotes it. The prawn does not moult while undergoing gonadal maturation or incubating the eggs.

Penaeid prawns are generally heterosexual. The female prawn is usually larger than the male. In the male, the endopodites of the first pair of pleopods are modified to form a copulatory organ, the petasma or andricum. The second pleopod also shows an accessory structure, the appendix masculina. In the female, the most striking character is the presence of a ventral thoracic structure, the thelycum, situated between the last three pairs of thoracic legs. In the female the opening of the genital ducts are situated on the bases of the coxae of the third pair peleopods, while in the male they are on the last pair.

Mature ovary of penaeid prawn is a paired organ, situated dorsally, extending from the base of the rostrum to the last abdominal segment. It is bilaterally symmetrical and partly fused. Each half of the ovary consists of three lobes. The anterior lobe is slender and runs along the anterior portion of the oesophagus and gastric mill. The middle lobe has 6 or 7 finger like lateral lobules which entirely fill the area between the anterior region and posterior border of the carapace. The thin oviducts start from the tips of the penultimate lobes of the middle lobe on either side and run downward to the external gonopore on the third pereopods. The posterior lobe extends the length of the abdomen. The two halves of the ovary are united by two commissures, one at the base of the anterior lobes and the other at the tip of the posterior lobes in the sixth abdominal segment. Thelycum is the modified sternal plates of the 4th and 5th thoracic segments and consists of an anterior plate, a pair of lateral plates and a posterior plate. The structure of the thelycum varies in different species.

The male reproductive system consists of a pair of testis, vas efferens, vas deferens, terminal ampoule and a petasma. The testis lie on the dorsal surface of hepatopancreas, ventral to the heart. Each half of the testis consists of several lobes extending over the surface of hepatopancreas. The vas efferens arises from the testis lobes and join into the proximal part of the vas deferens which is a thin delicate tube running downwards and backwards. The distal part of the vas deferens is thicker and has wider lumen. On the coxa of the last pair of throacic legs the vas deferens gets dialated and ends in a distal terminal ampoule. The ampoule is sac-like and primarily a glandular structure with a thick muscular coat and contains spermatophores and a white thicky fluid.

Petasma is the modified endopods of the first pleopods. Its structure and shape vary from species to species.

Spermatozoa of penaeid prawn are minute, usually globular or ovoid in shape with a very short tail. In *Parapenaeopsis stylifera* they are elongated and cylindrical.

Based on the external changes in colour, size and texture and microscopical examination of the ovary, five maturity stages have been recognized in the maturation of the ovary.

I-Immature Stage

Ovaries of immature prawns are thin, translucent unpigmented and confined to the abdomen. They contain oocytes and small spherical ova with clear cytoplasm and conspicuous nuclei.

II-Early Maturing Stage

The ovary is increasing in size and the anterior and middle lobes are developing. The dorsal surface is light yellow to yellowish green. Opaque yolk granules are formed in the cytoplasm and partly obscure the nuclei. The developing ova are clearly larger than the immature stock.

III-Late Maturing Stage

The ovary is light green and is visible through exoskeleton. The anterior and middle lobes are fully developed. The maturing ova are opaque due to the accumulation of yolk.

IV-Mature Stage

The ovary is dark green and clearly visible through exoskeleton. The ova are larger than in the preceeding stage and the peripheral region becomes transparent.

V-Spent-Recovering

After the extrusion of eggs, the gonad reverts almost immediately to the immature condition. This stage is therefore, distinguishable from that found in the immature virgin females only from the size of the prawn.

In males, the five maturity stages are:

Stage I

Testis lobes are not developed. Generative portion is present in the tubular portion upto vas deferens. Spermatozoa are not formed.

Stage II

Testis lobes are fairly developed. Generative tissue is very prominent in the tubular portion.

Stage III

Testis lobes are well developed. The two components of the tubular portion are well represented. Spermatozoa are present in the lumen of the follicles, tubular portion, and in the terminal ampoules.

Stage IV

Testis lobes are well developed. Spermatozoa present in the terminal ampoules. This is the fully mature stage.

Stage V

Testis lobes contain only spermatozoa but not the other stages of spermatogenesis.

The size at first maturity of males on the basis of fusion of two halves of petasmal endopodites is found to be 102 mm in *P. indicus,*

74 mm in *M. monoceros*, 71.6 mm in *M. affinis*, 53.6 mm in *M. dobsoni* and 59 mm in *P. stylifera*.

In the population, most of the commercial penaeid prawns show protracted breeding season often extending throughout the year. However, peak spawning season is found to vary from species to species from place to place, and from year to year. Usually most of the species show two peaks coinciding with the onset of the southwest monsoon and north east monsoon.

Spawning more than once in a season has been observed in a number of penaeid prawns. The individuals of *P. indicus, M. dobsoni, M. affinis* and *P. stylifera* are found to breed five or more times during their life span after attaining the first sexual maturity. After each spawning, the individual prawns again attain maturity within about 2 months.

All the commercial penaeid prawns of India breed in the sea in relatively deeper waters of the inshore ground. *M. dobsoni* and *P. stylifera* breed within 25 metre depth region, while the spawning grounds of the larger species, such as, *M. monoceros* and *P. indicus* are found to extend to still deeper waters upto 50-60 m depth zone. *M. brevicornis* in the Bombay region moves to deeper waters for spawning, whereas in Hooghly, it is found that mature females move from the upper to lower reaches of the estuary for spawning. *M. affinis. P. stylifera* and *P. maxillipedo* prefer areas of soft mud, rich plankton and shallow coastal waters for mating and spawning.

Fecundity and Spawning

Penaeid prawns have high fecundity. The number of eggs produces by female prawn varies from species to species and with the size of the prawn (Table 8).

During mating the males closely follow the females. Normally one male follows one female. The female moults while being followed by the male. This courting behaviour lasts for 3 to 7 minutes. Soon after moulting, the female lays her body sideways and jumps about in water bending her body ventrally when the male advances to the side of the female and embraces her on ventral side. The pair by keeping their ventral side close together swim about in the water, during which process the copulation takes place. During copulation, the male by means of petasma directs the spermatophores released through the openings on the fifth pair of legs to the thelycum of the

Table 8

Species	Size of First Maturity of Female (mm)	Fecundity Minimum	Fecundity Maximum	Peak Spawning Season
P. indicus	130.2	68000 (140mm)	731000 (200mm)	October-November; May–June
P. semisulcatus	23 (carapace length)	67900 (29 mm CL)	660900 (45 mm CL)	June–September; January–February
M. dobsoni	64.1	34000 (70 mm)	1,60,000 (120 mm)	April–August; October–December
M. affinis	88.6	88000 (95 mm)	3,63000 (160 mm)	October–December; January–March
P. stylifera	63.2	39500 (70 mm)	2,36000 (120 mm)	December–May; November–December
M. brevicornis	100			March–April
M. monoceros	–			July–August; November–December
P. sculptilis	75			December–May
P. hardwickii	–			October–July
M. kutchensis	–			February–September

female. The time spent on copulation is only 3 or 4 minutes. After copulation the male and female seperate.

Prawns spawn at night generally between 8 p.m. and midnight. The eggs are released while swimming in the columnar waters or near the bottom. During the process of releasing the eggs, they bend down the body posterior to the 4th abdominal segment and show side-wise movement. The fifth pair of legs are held tightly against the body. The eggs are dispersed by the movement of the pleopods. The time required for release of eggs is very short. In some of the species like *M. bennettae* spawning and mating are governed by lunar periodicity. Outward run takes place during new-moon cycle and spawning during full-moon phase.

All the commercially important penaeid prawns of India are known to breed and undergo early development in the open sea. Although these prawns enter the estuaries and backwaters in mysis and post larval stages and attain almost all the adult characters including the development of secondary sexual characters in this environment, it is believed that the maturation of ovary and subsequent spawning takes place only in the sea. Environmental factors such as high salinity, low bottom temperature and greater pressure of the deeper water of the sea play important role in this phenomenon, besides the physiological changes that trigger the maturation of the ovary. Intensive spawning activity is generally related to rise in water temperature. Water temperature around 28°–30° C and salinity of 28–34 ppt are found to be suitable for the maturation of the ovary and spawning.

Larval Development

The penaeid larvae pass through 6 nauplier stages, 3 protozoeal stages and a variable (3-7) number of mysis stages before becoming post-larvae.

Penaeus

Nauplius

The inner margin of the antennule has 2-3 setae, the distal one being very long.

Protozoea

The frontal organ is not overhung by pointed spine. Third maxilliped is absent in protozoea I, Protozoea II has supraorbital spines with bifid tip. The outermost seta on each furca of the telson is dorsally placed. Protozoea III has 8 pairs of furcal spines.

Mysis

Carapace with prominent rostrum, well developed supra–orbital and hepatic spines, the pterogostomian spine sharp, no antennal spine is present. Abdominal segment 4-6 with postero-dorsal median spines, the 3rd segment may also have a minute spine, the 5th and 6th segments with a pair of posterolateral spines. Uropod with the outer margin of exopod produced into a very prominent disto-lateral fixed spine beyond which the fringing setae are arranged, the outermost member of this series of setae is shorter than the distolateral spine and is non-plumose.

Post Larvae (P_1 and P_2)

Post larvae long and slender. Rostrum with one or two dorsal spines, supraorbital spine present, postero-dorsal and postero-lateral spines present on 5th and 6th abdominal segments. Telson with 8 pairs of spines. Scaphocerite long and narrow, broader anteriorly than proximally. Mandibular palp cylindrical with distal segment smaller than proximal segment.

Metapenaeus

Nauplius

The antennule has three short inner lateral setae, the size decreases gradually from the distal to the proximal seta.

Protozoea

Frontal organs overhung by pointed spine in protozoea I. Simple supra-orbital spines present in protozoea II and III. Caudal furcae short and broad, with shallow and moderate cleft and outermost pair of furcal spines slightly ventrally disposed, only 7 pairs of setae in protozoea-III.

Mysis

Carapace without supra orbital spine, but with antennal, pterygostomian and hepatic spines. Only abdominal segments 5 and 6 with dorsal spines, no lateral spines on any segment. On exopod of uropod the distolateral spine which is a continuation of the outer margin is absent in Mysis I and is very small in subsequent stages, being shorter than the outermost movable seta which is non-plumose.

Post Larva

Post larvae small. Rostrum short with 3 dorsal spines, no supra–orbital spine. Dorsal spine present only on 6th abdominal segment. Telson with 7 pairs of spines. Scaphocerite uniformly broad in distal and proximal halves. Mandibular palp club-shaped, distal segment broader than proximal segment.

Parapenaeopsis

Nauplius

The antennule with 3 inner setae, upto Nauplius III. The setae are short slightly decreasing in length from the distal to the proximal seta. But in nauplius IV to nauplius VI the proximal seta becomes thin and greatly elongated and sharply bent in distal one third.

Protozoea

Frontal organ not overhung by sharp spine. Rostrum short and straight, no supra-orbital spine. Postero-lateral spines on 6th abdominal segment inconspicuous in Protozoea III. Cleft in telson deep and wide, the caudal furcae narrow and long, the outermost pairs of spines laterally disposed and seperated from the penultimate pair by a wide gap. Antennules longer than antennae.

Mysis

Carapace with small supra-orbital spine at least in mysis I and II. Hepatic spine is absent in all mysis stages, antennal and pterygostomian spines are present. Abdominal segments 5 and 6 with dorsal spines. No lateral spines on 6th abdominal segment. Exopod of uropod lacks the fixed distolateral spine in all mysis stages.

Post Larva

Post larva (P_1 and P_2) is stout. Rostrum is short and curved with 3–4 spines. Supra–orbital spine is absent. No dorsal or lateral spines are present on abdominal segments 1-5. Telson with 8 pairs of spines and a posteromedian spine. Scaphocerite short and broad, broader proximally than distally. Antennal flagellum is longer than scaphocerite with more than 10 segments. Mandibular palp with a large oval distal segment and a smaller triangular proximal segment.

Bionomics of *Metapenaeus dobsoni*

Penaeid prawns, with the exception of *Metapenaeus monoceros* breed only in the sea. *Metapenaeus dobsoni* breed mainly in the sea. Eggs and larvae have been so far collected only in the sea, though early post-larvae attributed to *Metapenaeus monoceros* have been recorded from brackish water. The adult prawns though found throughout the year in brackish waters fails to show any larvae or eggs. While mature males with well developed spermotophores have been frequently observed in brackish water, no female either impregnated or with ripe ovary has so far been obtained from the same areas.

Sexual union and ripening of the female gonad take place only in the sea. Adult females from the sea have invariably the fifth pair of legs reduced to stumps and this degeneration of limbs in brackish water environment does not take place and is some way connected with reproduction. These facts strongly suggest the sea as the breeding place of this species.

The breeding period seems to be fairly long from September to April, with a peak from November to March. From September to the middle of January large numbers of eggs occur in plankton, after which the number drop abruptly. Plankton samples upto January contain mostly larvae of *M. dobsoni*. As such the period September to December or January forms the peak of the spawning season.

The off shore collections, between 24–26 m depth region, contain very few penaeid larvae and practically no eggs. A good number of breeding females liberate their eggs in the inshore waters upto or slight beyond 24-26 m. depth line. The Indian species spawns in the comparatively shallow inshore waters.

Impregnated females have been obtained practically throughout the year, the largest number being in the months of August to January and this period roughly coincides with the peak breeding season. Females soon after sexual union can be easily recognized because of the pair of prominent white pads that cover the surface of thelycum. Each pad is completely formed in the vas deferens and can be found together with the spermatophore in the terminal dialation of the vas deferens situated at the base of the last leg. These pads are rudimentary homologues of the appendages of the spermatophore of "Penaeus" the function of which is "for attachment of the exposed spermatophores in Penaeids with open thelycum. The smallest impregnated female measure over 65 mm.

The feeding habits of adult prawns are scavengers, but many are carnivorous and several species prey upon marine larvae and microscopic algae. The same is true for young ones as well. Small post-larvae thrive well on plankton containing a good proportion of algal constituents. Later on small quantities of finely minced clam or fish muscles are added to plankton. The food of 21-88 mm prawns in their natural habitat in general consists of varying amounts of organic matter mixed with sand and mud. Fragments or entire bodies of small animals and algae including diatoms compose the organic matter of which the proportion of vegetable constituents has been found to be less in the larger individuals. In prawns measuring 21-45 mm the stomachs were practically full of fragments of the alga, *Cladophora* mixed with mud and some diatoms. Individuals measuring 65-75 mm contained only a few fragments of the same alga. The animal matter largely consists of entirely or partly digested bodies of Foraminifera, Copepoda, Nematoda, Amphipoda and Gastropoda.

Besides Cladophora, a number of diatoms are frequently present, the more common of them are *Fragilaria, Coscinodiscus, Pleurosigma, Navicula* and *Cyclotella*.

Male post larva of 3.5 mm when reared into glass troughs and fed regularly with calm meat for 2 months and 20 days grow to 47 and 41 mm respectively. With finely minced fish muscle they grow 38 mm in two and half months and 43 mm in 5 months. Dominant size group from the commercial catches are those between 60 and 80 mm which could not be older than 7 or 8 months. During the breeding period females of 50-70 mm were found with fully ripe ovaries so that sexual maturity may also be attained at about this age.

The life cycle of each individual is completed in two types of environment, namely, the brackish water of estuaries and lagoon connected with the sea and the sea itself. The entire course of larval development and the earliest post-larval phase are passed through the open sea. Post-larval penaeids enter back water when they are 1-2 cm in length. Soon after the moult from first post-larval stage the young ones cease to be pelagic. The smallest captured from brackish water at various localities are 15-25 mm in length. Prawns of this size are caught in fair numbers from the foreshore waters upto a depth of 8 m in November and December. It is quite probable that migration from the sea taken place when the young have reached this size and that successive batches pass out of the sea during the greater part of the breeding period. The maximum size of prawns collected from backwaters and estuaries does not exceed 80 mm. The backward migration into the sea may take place soon after the monsoon rains start, the prawns being carried in the current flowing into the sea. As the largest individuals measure as much as 120 mm and have been found only in the sea they do not seem to leave the sea afterwards and re-enter their former habitat. Osmotic regulation in penaeid prawns has shown their high adaptability to water of varying salinity among migratory penaeids.

The vast majority of *M. dobsoni* population do not grow beyond 60-65 mm in length in backwaters. Those growing beyond 70 mm are quite few and have seldom formed more than 2.5 per cent. Even those belonging to 66-70 mm group make up only a small proportion.

The breeding period of the species along the Malabar coast is fairly long extending from September to March or April. The species has been found to grow in the marine environment during the first, second and third year of their lives as below.

Sex	1st Year	2nd Year	3rd Year
Male	About 70 mm	About 90-95 mm	About 110 mm
Female	About 75-80 mm	About 100-105 mm	About 120 mm

The largest female obtained so far measured 124 mm, while males measured only 111 mm, though an exceptional specimen caught at West Hill was as much as 118 mm in length. The percentage of males and females measuring 100 mm and above in the total number of each sex measured and calculated as 7.5 per cent and 12.1 per cent respectively.

The proportion of females is slightly higher (52.9 per cent in 1952, 51.3 per cent in 1953 and 51.9 per cent in 1950-51) than that of males (47.1 per cent in 1952, 48.7 per cent in 1953 and 48.1 per cent in 1950-51) in marine environment.

The peak period of the breeding of the species is August to December. Migration of the species commences when they are in second stage and quite a large proportion passes into such brackish water areas before they reach a length of about 7 mm. Throughout the breeding season, therefore, immense numbers of these fry would be continuously passing into these waters from the sea.

M. dobsoni ranks high among the commercially important species of penaeid prawns occurring along the south-west coast of India contributing a good portion of the catches practically throughout the year. *M. dobsoni* is the dominant species in the prawn fauna of the brackish waters in all but three months (June, July and August) when there is high floods due to monsoon rains. During monsoon months, the brackish water canal becomes practically fresh and in consequence, prawns are quite scarce.

Marine fishery of *M. dobsoni* is largely seasonal from about April to October. Fishing is done usually in 16-20 metres depth of water unless large concentrations are discovered nearer the shore. The net used is the boat seine. Cast nets are also used if the prawns occur quite close to the shore.

Chapter 6
Prawn Fishery in Indian Coasts

The prawn fishery of India is both dynamic and complex, supported by multiple species that coexist in the same fishing grounds. The commercial penaeid prawns are subjected to exploitation in the juvenile and adult phases of their life in two different ecosystems of estuaries or backwaters and sea. Wide fluctuations in the catch are observed in all the regions. The production of shrimp in India during 2006 through capture was 3.43 lakh tonnes.

Nevertheless their biological features such as, their capacity to produce large number of eggs, protracted breeding season, faster rate of growth, short life span and their ability to withstand wide environmental changes help to maintain their population. But the exploitation of prawn resources from the wild stock in recent years, increased so rapidly that in certain areas of our coast, it has almost reached or has already reached the optimum level.

Prawns and shrimps form the most economically significant group in the marine and brackish water fishery resources. In the freshwaters, however, they contribute to only a subsistence fishery except in certain regions. About 55 species of prawns and shrimps that are either commercially exploited at present or have great commercial potentials occur in the Indian waters Accounting for an average catch of 0.14 million tonnes, this group represents about 13 per cent of the total marine fish production of the country. Substantial quantities of prawns and shrimps, estimated at about one third of the marine prawn landings are also caught from estuaries and backwaters.

Till about five and a half decades back, prawns used to be sent to neighbouring countries in dried or cured form earning an annual revenue of a few million rupees. From this, today prawns have assumed an important place especially as a commodity supporting an export trade of sizeable magnitude. Prawns alone exported from India in 2006 was 612641 tonnes valued at about 8363.53 crores rupees.

The prawn fishery consists of penaeid and non-penaeid prawns. The penaeid prawns which constitute about 62 percent of the total prawn landings are the commercially important ones. The fishery is supported by multiple species that co-exist in the fishing grounds and is characterized by wide seasonal and annual fluctuations in abundance. One striking feature of the prawn fishery is that most of the penaeid prawns are subjected to exploitation in the juvenile phase. In terms of landings, Maharashtra ranks first with 47.5 percent in prawn production followed by Kerala (30.7 per cent), Andhra Pradesh (5.5 per cent), Tamil Nadu (4.6 per cent), Gujrat (4.1 per cent), Karnataka (3.8 per cent) and West Bengal and Orissa (3.1 per cent). Although Maharashtra ranks first, the highest catch of penaeid prawns which are exported comes from Kerala. Until a few years back, the prawn fishery was concentrated by and large on the west coast of India. But in recent years exploratory surveys carried out have revealed existence of commercially exploitable stocks of good sized prawns along the east coast, which has led to substantial increase in fishing along the coast. Exploratory surveys carried out along the continental shelf edge and slope of the south-west and south-east coasts have located potentially rich fishing grounds of deep water prawns. These grounds on the south-west coast are about 5000 sq. km. in extent and the magnitude of the harvestable resource from this area has been estimated to be about 5,300 tonnes per year. These resources are, however, not-been fully utilized because of distant location of the fishing ground and the limited demand for the varieties of prawns available there.

There has been a steady increase in the catch of penaeid and non-penaeid prawns which after reaching a peak in 1975 (2,20,751 tonnes) have shown a decline. There is also a steady decrease in the size of the prawns as well as catch per unit effort. Due to the entry of a large proportion of small sized prawns in the export trade, there has been a tendency for decrease in the unit value realised from export.

There are definite indications of over fishing of prawns and that suitable conservation measures will have to be enforced which may include restrictions on fishery of juveniles, even though there is a risk of the economy of the traditional fishing sector being seriously affected.

Capture fisheries is becoming more and more capital intensive especially with the rising coast of fuel and energy. Although culture fisheries can not replace capture fisheries, they would certainly supplement and augment the total yield. Prawn culture was a neglected field in India; and formed one of the constituents of the brackish water culture, practised traditionally in Kerala and West Bengal. Here the fields are impounded by bunds, source of prawn seed is from the wild stock brought in by the incoming tidal currents. Culture practice neither involves any husbandry nor management. Prawns are allowed to grow in the fields for a short periods feeding on the natural food available in the environment. Thus the success or failure of the operation depends largely on the availability of seeds in the nature and the biotic and abiotic conditions of the field. The production of prawn from this field is very low varying from 500-900 kg/ha/year.

The major cultivable species of penaeid prawns in Indian waters are *Penaeus monodon, P. merguiensis* and *P. indicus*. Though *P. monodon* is widely distributed, marine landings indicate that *P. merguiensis* and *P. indicus* are far more abundant than *P. monodon*. The post-larval abundance of *P. monodon* is very much less than that of either *P. merguiensis* or *P. indicus*. During peak periods of post-larval ingress, *P. indicus* constituted 80-90 per cent of the collections. It was estimated that upto 10,000 post-larvae of *P. monodon* could be segregated per tide per day. During April, 1977 it was possible to segregate 32,000 post-larvae of *P. monodon* in 90 man-hours. Quantitative estimates of post-larval and juvenile abundance of other prawn species are lacking. While natural seed abundance generally fluctuates from season to season and from place to place, the information presently available clearly indicate for organising systematic collection of prawn seed from natural sources as an industry. That has already been done and the seed was made available to interested shrimp farmers.

All the species of prawns are not using the brackish water environments as their nursery grounds. Considerable variations occur among the prawns both in degree to which the brackish water environment is put to use by each species in its life history and in the distribution of the parents and juveniles population in the marine brackish environments.

The estuarine, back water and mangrove environments play a vital role in the life cycle of prawns and in the establishment of its fishery. The correlation between certain penaeid species and mangrove forests are "no mangroves"; "no prawns".

Table 9: Percentage Composition of Juveniles of Penaeid Prawns Among Porto Novo Biotopes

Species	Size Range in mm	Aquatic Biotopes		
		Estuary	Back Water	Mangrove
Penaeus indicus	20-120	21.9%	33.6%	44.9%
Penaeus semisulcatus	15-115	46.9%	25.4%	2.0%
Metapenaeus dobsoni	10-50	12.7%	6.8%	32.5%
Penaeus monodon	25-115	6.3%	10.3%	6.2%
Metapenaeus monoceros	12-80	5.7%	11.4%	9.3%
Metapenaeus brevicornis	10-46	3.3%	3.9%	3.8%
Penaeus merguiensis	20-96	3.2%	8.6%	1.3%

Prawn Fishery Along the North-East Coast

The traditional fishery for prawns along the north-east coast with different varieties of shore seines and boat seines had been restricted to a narrow strip of coastal waters. Commercial trawling for prawns by small mechanized boats of 10-11 m length with 45-70 HP diesel engines started towards the end of 1967 at Visakhapatnam. In 1972 two 24m Gulf of Mexico type trawlers successfully started fishing for prawns. The big trawlers of 22-25 OAL are made up of steel with 380-450 HP. diesel engines. The large trawlers have freezing facilities on board the vessel and can stay at sea for 18-23 days.

With the aim to catch penaeid prawns these vessels operate two identical nets simultaneously from the out triggers on both sides of the boat. Generally a haul lasts for about 3-4 hours, although hauls upto 6 hours duration are quite common. These trawlers generally operate between Pentakota (17°N lat.) in the south and Sunderbans (21°N lat.) in the north including the vast areas of the Sandheads. Although they fish over wider areas, most of the effort is expended between Gopalpur (19°N lat.) and the Sunderbans (21° N lat.). These vessels generally fish in the depth range of 40-80 m and rarely in 10-40 m and beyond 80 m. depth zones.

From the available fishing logs of 13 vessels in 1983-84, 38 vessels in 1984-85, 43 vessels in 1985-86 and 45 vessels in 1986-87, fishing hour or trawling hour, the time actually spent in trawling was taken as a standard unit of effort and the catch obtained in one hour of trawling or catch per trawling hour (CPH) was considered as an index of abundance.

The species composition of different categories of prawn catch were "Tigers" (*Penaeus monodon, P. semisulcatus* and *P. japonicus*)/ "Whites" (*Penaeus indicus, P. merguiensis,* and *P. penicillatus*); "Browns" (*Metapenaeus monoceros, M. ensis* and *M. affinis*) and "Others" (*M. brevicornis* and *M. dobsoni*).

The fishing effort was at a minimum during April–June, from when gradually increased to reach a peak during October–December and then declined gradually to March. This trend was followed in more or less all the four years.

In 1983-84, 55 vessels landed 2353 tonnes of prawn by trawling for about 143961 hours (CPH of 16.3 kg). Prawn landings varied from 10.8 tonnes in May to 395.5 tonnes in January, while the CPH varied from 5.9 kg in April to 27.2 kg in October. Almost 55 per cent of annual catch was landed in October–January.

In 1984-85, 60 vessels land 2815 tonnes of prawn with an effort for 139929 trawling hours (CPH of 20.1 kg) Prawn landings gradually increased from 1.8 tonnes in May to 431.2 tonnes in September and then declined to 64.9 tonnes in March, while the CPH increased from 5.2 kg in May to 28.5 kg in November and then declined to 8.0 kg in March.

The annual prawn landings in 1985-86 were at 3043 tonnes for an effort of 184874 trawling hours expended by 75 vessels. Prawn landings gradually increased from May (20.3 tonnes) till November (471.0 tonnes) and then declined gradually till May; whereas CPH gradually increased from July (15.7 kg) to September (21.7 kg) and then declined gradually till February (9.2 kg)

In 1986-87, 100 vessels landed 3077 tonnes of prawn for an effort of 195300 trawling hours with an average CPH of 15.8 kg. Prawn landings gradually increased from June (45.8 tonnes) till December (558.9 tonnes) and then declined till March (45.8 tonnes); while the CPH increased from 16.1 kg in August to 19.8 kg in October and then declined gradually to 7.6 kg in March.

The prawn catch in all the 4 years was composed of 'tigers' (8.1 per cent); 'whites' (29.9 per cent); 'browns' (60.7 per cent) and others (1.3 per cent). The portion of others in the first three years were negligible. It is observed that 'browns' dominated the catches in almost all the months followed by "whites" and 'tigers'.

The landings of tigers in 1983-84 were estimated at 194.3 tonnes, forming about 8.3 per cent of the annual prawn landings. The landings gradually increased from May (1.4 tonnes) till November (34.3 tonnes) and then decreased in December and January followed by an increasing trend till March (20.4 tonnes) CPH also indicated more or less similar trend with the annual CPH estimated at 1.4 kg.

With an estimated catch of 236.5 tonnes, 'tigers' formed about 8.4 per cent of the prawn landings in 1984-85. The landings exhibited peaks in September (43.8 tonnes), November (37.7 tonnes) and February (23.0 tonnes); whereas peaks for CPH were observed in August (2.7 kg), November (2.1 kg) and February (1.5 kg). The annual average CPH was at 1.7 kg.

233.5 tonnes of 'tigers' forming about 7.7 per cent of the prawn landings were recorded in 1985-86. The peak landings were observed in August (30.4 tonnes), October (33.7 tonnes), January (32.6 tonnes) and March (19.9 tonnes). The CPH also followed the same trend and the average annual CPH was 1.3 kg.

A catch of 255.4 tonnes 'tigers' formed about 8.3 per cent of the prawn landings in 1986-87. The landings exhibited peaks in August (48.8 tonnes), October (41.9 tonnes) and December (38.0 tonnes); while the CPH gradually declined from August (2.3 kg) till February (0.4 kg).

Annual landings of whites (*P. indicus, P. merguiensis* and *P. penicillatus*) varied from 719.7 tonnes in 1983-84 to 954.4 tonnes in 1986-87; while the CPH varied from 4.8 kg in 1985-86 to 5.8 kg in 1984-85. During 1983-84, the landings for whites exhibited peaks in August (53.6 tonnes), October (121.6 tonnes) and January (154.0 tonnes); whereas the peaks in CPH were in June (5.3 kg), October (10.0 kg) and January (7.4 kgs). Whites formed about 30.6 per cent of the prawn landings in 1983-84.

With a catch of 889.6 tonnes, whites formed about 29.2 per cent of the prawn landings in 1985-86. The landings gradually increased from April (0.8 tonnes) and reached a peak in November (172.2 tonnes) and then declined thereafter till March (53.3 tonnes). The CPH gradually increased from 3.9 kg in July to 7.7 kg in October and then declined till March (2.6 kg).

The catch of whites was estimated at 954.4 tonnes forming about 31 per cent of the prawn landings in 1986-87. The landings increased from June (13.5 tonnes) till October (239.1 tonnes) and then declined gradually till March (2.2 tonnes). The CPH showed random variations in June-September and declined from October (8.8 kg) till March (0.2 kg)

Pooled data for the four year period indicate that October-December was the best period for the fishery of whites with better landings and CPH. The landings gradually increased from April till October and then declined gradually till March.

With an estimated catch of 1417.4 tonnes of browns (*Metapenaeus monoceros, M. ensis,* and *M. affinis*) formed 60.2 percent of prawn landings in 1983-84. The landings and CPH indicated peaks in September, January and March. Annual average CPH was 9.9 kg with variations from 3.7 kg in April to 18.1 kg in September.

1757 tonnes of 'browns" forming about 62.4 per cent of the prawn landings were caught in 1984-85. The landings and CPH exhibited peaks in September, November and January. CPH varied from 4.9 kg in May to 20.0 kg in November with the annual average estimated at 12.6 kg.

A catch of 1912.8 tonnes of 'browns' formed about 62.9 per cent of the prawn landings in 1985-86. The landings more or less gradually increased from 10.2 tonnes in May to 294.7 tonnes in

January. The CPH varied between 5.2 kg in February to 14.6 kg in September with the annual average at 10.4 kg.

In 1986-87, 1763 tonnes of 'browns' were landed forming about 57.3 per cent of prawn landings. The landings as well as CPH declined as compared to 1985-86. The landings varied from 21.1 tonnes in April to 356 tonnes in December. The CPH varied from 6 kg in March to 13.4 kg in September with the average estimated at 9.0 kg.

Pooled averages for the four-year period indicate that August–January was the best period for the fishery of 'browns' with about 78 per cent of the catch landed during this period. The CPH was also very high during the period as compared to February–July period (Table 10).

Monthly abundance of total prawns off Kalingapatnam varied (CPH) from 1.96 kg in February to 27.69 kg in September. The CPH varied from 4.44 kg in February to 15.26 kg in October off Gopalpur. In many areas high CPH was recorded during August–December period and moderate values in June–July and poor values in February–May.

The abundance of tigers was more in the grounds off Balasore in July; off Kalingapatnam, in Sandheads II and Off Anchorage in August and off Gopalpur, off Chilka lake, off Konarak and off Paradeep in October. Generally the abundance of tigers was less in most of the areas during March-June period.

The abundance of 'whites' was more off Kalingapatnam and off Paradeep in August and September; while it was more off Gopalpur, Sandheads II, off Balasore and off Anchorage in October; off Sandheads I and off Sunderbans in November and December. CPH was poor in most of the areas during January–June period.

'Browns' were abundant in July and September off Kalingapatnam; in June off Gopalpur; in June, December and March off Chilka lake; in August and October off Konarak; in July and January off Paradeep; in July-October in Sandheads I; in July to February in Sandheads–II; in July, September and January off Balasore; in August-March off Anchorage and in November-January off Sunderbans.

The abundance of prawns randomly fluctuated in 11-40 m depth zone between 10.83 and 15.46 kg; while it gradually increased

Table 10: Abundance of Prawns in Different Areas of North-East Coast of India

Area	Latitude °N	Longitude °E	Prawn Catch (kg)	CPH (kg)	Tiger Catch (kg)	Tiger CPH (kg)	White Catch (kg)	White CPH (kg)	Brown Catch (kg)	Brown CPH (kg)
Off Kalingapatnam	18°	84°	1865	12.73	116	0.79	656	4.48	1093	7.46
Off Gopalpur	19°	84°	2012	13.12	141	0.92	1262	8.23	609	3.97
Off Chilka	19°	85°	18759	12.63	2248	1.51	6981	4.70	9550	6.43
Off Konarak	19°	86°	3515	12.03	305	1.04	518	1.77	2692	9.21
Off Paradeep	20°	86°	3531	10.86	528	1.62	1173	3.61	1830	5.63
Sandheads I	20°	87°	12320	10.63	1913	1.65	4348	3.75	6059	5.23
Sandheads II	20°	88°	39653	11.99	2104	0.64	9816	2.97	27733	8.38
Off Balasore	21°	87°	24280	12.55	1518	0.78	11645	6.02	11117	5.74
Off Anchorage	21°	88°	15024	12.65	679	0.57	1966	1.65	12379	10.42
Off Sundarbans	21°	89°	6613	14.84	91	0.20	1194	2.68	5328	11.96

from 41-50 m (10.32 kg) to 91-100 m (16.36 kg). The abundance of tigers as indicated by CPH gradually increased from 11-60 m and then gradually declined beyond this depth. The proportion of 'tigers' in the total prawn catches also indicated a similar trend. The abundance of 'whites' was better in 11-40 m depth and then declined gradually beyond this zone. The proportion of 'whites' gradually declined from 11-20 m (89.66 per cent) to 91-100 m (2.32 per cent). The abundance of 'browns' gradually increased from 11-20 m (1.39 kg) to 91-100 m (15.3 kg). Similarly the proportion of "browns" in the prawn catches also increased gradually from 11-20 m (8.77 per cent) to 91-100 m (93.76 per cent).

The abundance of total prawns was more in November and December in 11-20 m depth range, September-December in 21-30 m, May-November in 31-40 m, May to October in 41-50 m, May-November in 51-60 m, August–December in 61–70 m, July–January in 71-80 m, August–January in 81-90m and August–December in 91-100 m depth ranges. In general the abundance was less in February–April in all the depths.

The abundance of tigers was more in October in 11-20 m, August–September in 21-30 m, May–October in 31–40 m, July–March in 41–50 m, May–August and December–February in 51–60 m, January, August and October in 61–70 m, and July–December in 71–80 m depth ranges. In 81–90 m and 91–100 m it was negligible in the catches.

The abundance of 'browns' was more in August–October in 21–30 m, May–June in 31–40 m, April–July in 41–50 m and May–July in 51–60 m depth ranges. In the depth range of 21–60 m 'browns' were abundant in almost all the months.

Prawn Fishery Along the South-West Coast

The two major prawn fishing grounds off Tinnevelly coast are at Manappad, with a rich seasonal fishery dominated by *Penaeus indicus* and at Punnaikkayal, Tuticorin, which presents a round–the–year fishery dominated by *P. semisulcatus*.

The near shore region of the continental shelf of south-east coast of India is well known for its rich corals, and the bottom is generally rocky and sandy. But there are narrow gullies of muddy bottom with rich concentrations of prawns, about 6 km away from the shore at Manappad and slightly closer at Uvari fishing village. Fishing is carried out usually at depths of 15 to 25 m, occasionally extending up to 30 to 32 m. The muddy nature of the bottom in this region makes trawling possible.

Prawn fishery off Manappad is invariably seasonal, unlike that of Punnaikkayal where exploitation is done throughout the year. The fishing season of Manappad by mechanised boats commences by June-July and terminates by October-November of each year. It is generally seen that the close of prawn fishing season at Manappad coincides with the commencement of the peak season further north in the Gulf of Mannar, at Mandapam and Rameswaram in the months of January to March.

P. indicus is the predominant species contributing to about 63.4 per cent in the grounds off Manappad. The other species represented in the catch are *P. semisulcatus* and *Parapenaeopsis stylifera*. The peak landings of *P. indicus* are July and August. Larger quantities of *P. stylifera* are landed towards the latter half of the season. The maximum landings of this species from Manappad is in the month of October. *P. indicus* formed only about 24.4 per cent of the total catch in December and January, when the catches are poor.

The catch rate of *P. indicus* from Manappad grounds was high in June and July (3.2 kg and 3.4 kg per hour respectively), but showed a steep decline in the following months. On the contrary, the catch of *P. indicus* from Punnaikkayal did not show any sudden fluctuation, and the fishery is more or less steady throughout the year with an estimated average catch rate of about 1.42 kg per hour.

P. indicus ranged in length (from tip of the rostrum to tip of telson) from 121 to 200 mm in males and from 131 to 215 mm in females. However, only 2.7 per cent of the males landed during the season belong to the size group 121-135 mm and 2.6 per cent of females in size group 131-140 mm. The dominant mode for males shifted from 151-155 mm in June to 156-160 mm in July and August and 171-175 mm in September and October. In the case of females, the dominant mode shifted from 171-175 mm in June to 151-155 mm in July, 176-180 mm in August, 186-190 mm in September and 196-200 mm in October. The recruitment to Manappad fishery is only in June and July for which the catch rates are high in these months.

P. indicus landed by mechanized boats from the adjacent fishing grounds off Punnaikkayal range in length from 96-200 mm in case of males and from 101 to 210 mm in case of females. The dominant modes of males and females from January to December showed a bigger size groups of 161-165 mm and 166-170 mm in males and 176-180 mm and 186-190 mm in females in August and September when similar sizes appeared at Manappad also. In most of the other months the species was represented mainly by smaller size groups only (126-130 mm, 121-125 mm, 136-140 mm and 141-145 mm in case of males and 126-130 mm, 121-125 mm 146-150 mm and 136-140 mm in case of females in November, December February and May respectively). Thus the recruitment of *P. indicus* into the fishery of Punnaikkayal fishing ground and nearby areas seems to be in the months of November–February and May as against June-July at Manappad.

A slight predominances of females over males is noticed from the fishing grounds off Manappad and Punnaikkayal with respect to *P. indicus*. The yearly average percentage of males and females from the former ground was 47.4 and 52.6 and from the latter ground was 46.7 and 53.3.

Late maturing and mature females are quite abundant in July, August, September and October, especially in September when they contributed 87.13 per cent. The presence of large numbers of maturing and mature individuals together with small number of impregnated individuals and negligible numbers of immature ones throughout the season would indicate that Manappad is spawning ground of *P. indicus*. The fishing grounds off Punnaikkayal also show that maximum number of maturing and mature individuals occur during the same period. This would also lead to the inference that the peak spawning season of *P. indicus* of this region is during July to October.

P. indicus breeds throughout the year, with two peaks, one in December-January and the other in May–June along the south-west coast of India. But the breeding activity of this species was pronounced in March and May to September at Chennai. There may be difference in breeding season of the species occuring along west and east coasts of India. At Punnaikkayal large numbers of immature prawns are landed in November (66.67 per cent), December (83.9 per cent), February (50.52 per cent) and May (56.86 per cent). At the same time females of late maturing, mature and spent-recovering were less during these months. So the recruitment of *P. indicus* into the fishery appears to be during the months of November to February and May.

P. indicus spends its juvenile stages in estuaries and move out into the sea after reaching a size of about 120 to 130 mm. The nearby Punnaikkayal estuary is supporting the fishery of its adjacent area, where the fishery is not seasonal and the landings are comparatively steady throughout the year ruling out any possibility of a major recruitment to the fishery off Manappad in May–July. Since at no time younger specimens are present in the fishery off Manappad, the stock produced there itself is ruled out. The prawn fishery of areas further north like Erwadi and Mandapam is mostly contributed by *P. semisulcatus* Thus recruitment of *P. indicus* from northern side is not likely. On the contrary, *P. indicus* form the dominant species of the fishery towards the southern side of Manappad. The prawn fishery of Kanyakumari district is exclusively constituted by large-sized *P. indicus*. The fishing season at Kanyakumari starts with the onset of the south-west monsoon in May–June and extends upto September–October. Within the region itself, the fishing season begins and also ends earlier in the northern areas. The significance of this fishery is that it occurs during the monsoon, when fishing activities all along the west coast are either stopped or very slack. Significantly the fishery off Manappad is exactly during the same period and contributed by similar sizes of prawns of the same species.

During the phenomenon of upwelling along the south-west coast of India, during south-west monsoon period fishes are forced to move deep waters or press very close to the shore during such physico-chemical disturbances. In the fishery off Cochin, all the important species other than *Metapenaeus dobsoni* move to deeper waters during this upwelling period in the monsoon season. The occurrence of *M. dobsoni,* larvae all the year round in the inshore waters and the scarcity of larval forms of *P. indicus, M. affinis* and *M. monoceros* in the same area from February to August indicate that it might be due to some migration of *P. indicus, M. affinis* and *M. monsceros* from the area. The rich fishery of large-sized *P. indicus* along the south-west coast of Kanyakumari and adjacent south-east coast including Manappad takes place at about the same period. The only possible source of recruitment of *P. indicus* to Kanyamkumari and Manappad is from the northern region, namely the Kerala coast with several estuarine nursery grounds where the species exists as a fishery. Hence it is probable that during the upwelling period of south-west monsoon *P. indicus* moves to the south and appears in the fishery of Kanyakumari district and also the adjacent areas on the east coast at Manappad.

The 30 m. depth contour line off Manappad is as broad as that off the Kerala coast which harbours a rich fishery of *P. indicus.* On the contrary the 30 m contour line off Kanyakumari district is extremely narrow. Hence, in all probability the broad and shallow shelf region with its muddy substratum, off Manappad and adjacent areas on the east coast may be providing a highly favourable habitat for the adults of *P. indicus* which at the time of physico-chemical disturbances during south-west monsoon are forced to move from the inshore waters of Kerala towards the south and on reaching the narrow stretch of the inshore region of Kanyakumari, moves further to the shallow spread of the shelf region on the south-east coast and gets recruited in the fishery there.

The peak season of abundance (November-January) of *Penaeus semisulcatus* coincides with the North East monsoon and the post-monsoon period. The influence of climatic conditions on the abundance of prawns has been observed, the prawn catches appeared to follow the pattern of rainfall, the landings being more with heavier rainfall in the case of stake-net catches from the backwaters of Cochin. The prawn fishery have been found to reach the maximum during the subsequent months followed by more than 80 per cent of the annual rainfall during June-October on the south-west coast of India.

Juvenile Prawn Fishery in Tinnevelly Coast, Tamil Nadu

Penaeus semisulcatus enjoy wide distribution along the coastal waters of India. It contributes to a major fishery and dominates the commercial catches along the southeast coast. The species being growing to a large size, is in good demand in the export industry and is increasingly exploited by mechanized boats.

The fishery of *P. semisulcatus* is limited to shallow coastal waters extending to about 20 km from Pattanamarudur to Tuticorin. It is also exploited occasionally at the Harbour Point (Tuticorin), Hare Island and Pattanamarudur. The substratum of the fishing ground is sandy, with corals and rocky patches. The ground slopes gently and is covered with a thick growth of marine plants, sea grasses and algae.

The inshore area is very shallow, the 6 metre contour being conspicuously broad from Vaippar to Tuticorin. It reaches the maximum width at Pattanamarudur and Tarravarculam, to about 6 to 7 km away from the shore. Most of the commercial penaeid prawns are known to be closely associated with shallow brackish water environment after their postlarval stages, showing a preference to shallow waters during their juvenile stages. The extensive shallow water area, though not brackish, along the Tinnevelly coast is one of the favourable factors for the occurrence of the juvenile population of the species here.

Traditional gear, small shore-seine type, with the help of Tuticorin type of boat with less man power is operated invariably in shallow waters of near shore region going upto a maximum distance of 0.75 to 1 km and a maximum depth of 2 to 3.5 meters. Operation of this gear is similar to that of a shore-seine but the haul is completed in 60 to 80 minutes by 3 to 5 people including women and children. Four to six hauls are made during 8 hours constituting a days work.

About 75 per cent of the total catch were constituted by fishes. *P. semisulcatus* form the entire bulk of the prawn catches. Stray occurrence of juveniles of *P. indicus, P. latisulcatus, Metpenaeus moyebi* and *Metapenaeopsis stridulans* were noticed occasionally. Nearly 60 per cent of *P. semisulcatus* caught were between 55-95 mm in length. The fishery is exploited round the year and catches seem to follow a more or less regular pattern rising to a peak in November-January. This peak coincided with the north east monsoon and post monsoon period. Catches were good in the months of June and July also. The catch per unit effort averaging to about 5.82 kg for the year rise to 11.33 kg at the time of peak landings. The total estimated catches during October 1978 to September 1979 was about 61 tonnes consisting of an estimated 79,78,660 juvenile prawns.

The size of prawns in the catches ranged from 36-40 mm to 136-140 mm in the case of males and from 36-40 mm to 146-150 mm in the case of females. However the fishery was supported mainly by prawns measuring 56-60 mm to 106-110 mm Recruitment of the younger size groups into the fishery took place throughout the year. But the entry of younger sizes ranging in modal length from 56-60 mm to 89-90 mm into the fishery was conspicuous in the months of March, April and May indicating this to be the main recruitment season. Entry of waves of smaller sizes was conspicuous in the months of September and December-January also. The modal lengths were noticed at the maximum in the months of October and November.

The sexes are not equally distributed in the fishery. There was a clear predominance of females in almost all the months constituting on an average to about 56.5 per cent of the total catch. Similar cases of female predominance in several commercially important prawn species were noticed during a study of the offshore prawn fishery of Cochin.

Limiting the fishery of the small juvenile prawn from shallow inshore and brackish water areas including estuaries will allow most of the juveniles of return to the offshore waters, where, they can be caught at bigger sizes.

Penaeid prawns in their juvenile stage are extensively fished from the shallow brackish waters and estuaries all along the coast of India. Considerable variation occur both in the degree to which the brackish water environment, the commercial penaeids utilise and in the distribution of the parent and juvenile populations in the brackish and marine environment. The non-entry of *Parapenaeopsis stylifera* into the brackish water environment was cited as an exception. Although juveniles of *P. semisulcatus* are found in estuaries in other areas, an exception to the set pattern of life history of the species is noticed in the fishery of the shallow coastal waters of Tuticorin which provide a purely marine environment.

Occurrence of juveniles of *P. semisulcatus* in marine habitat with thick growth of aquatic plant has been noticed in Durban Bay on the east coast of Africa, where the occurrence of this species in large numbers were associated with good growth of eel grass (*Zostera capensis*) and relatively scarce when the growth of eel grass was poor. The juveniles of *P. semisulcatus* measuring 3.2 mm to 17.0 mm in carapace length spent their life from late August to middle to October in areas of the Seto Island Sea of Japan where *Zostera marina* are growing. The abundance of juveniles of *P. semisulcatus* in the shallow inshore waters of Tinnevelly coast rich with marine plants and their migration as they grow in size to deeper water with muddy substratum is in agreement with the observations in Seto Island Sea. But the species was observed to form significant portion of the prawn catches in muddy brackish water areas (Bheris) of West Bengal and the juveniles of this species are often well represented in the brackish water fishery of the west coast of India.

A direct relationship between *P. semisulcatus* abundance and rain fall have been observed in back waters of Cochin (heavier the rainfall more the stake net catches), in the Texas coast, in the Mangalore estuary on the south west coast of India and along the Tinnevelly coast, the peak season of abundance (November-January) of the species coincides with the northeast monsoon and the post monsoon period.

Andaman and Nicobar Islands

Penaeus canaliculatus is distributed in Red Sea, South Africa, Mauritius, Sri Lanka, Andaman Sea and Pacific Ocean. The telson bears lateral spines which are distinguishing characters by which the species can be seperated from *Penaeus japonicus* which has also more or less similar colouration while alive. The rostrum is with only one lower tooth as in *P. japonicus*. The thelycum also is characteristic in shape and structure of the anterior plate. Male of 10.5 mm length and females of 19.0 to 19.5 mm length have been obtained from trawl net catch at Marine Bay of Andaman Island at 2-12 m depth.

Two females *Penaeus merguiensis* of 27.0 mm and 36.1 mm length was obtained in trawl net catch at Marine Bay, Andaman Island. The species is also distributed in West Pakistan, south west coast of India, Sri Lanka, east coast of India, Mergui Archipelago, Singapore, Hong Kong, Philippines, East Indies, Japan and Australia. The species can be easily distinguished from *P. indicus* and *P. penicillatus* by the presence of the characteristic deltoid crest at the base of the rostrum which is usually reddish in colour with darker margins in adult specimens. The gastro-orbital carina is clearly defined in adults.

Three male specimens of *Metapenaeus dobsoni* were recorded from trawl catch at Marine Bay, in Andaman Island. The species is also found in Gulf of Suez, west coast of India, Sri Lanka, east coast of India, Malaya, Hong Kong, Japan and Australia. The free filaments of the distomedian projections of petasma on the dorsal aspect are well developed in the adult specimens. The impregnated females have conjoined white pads on the thelycum.

Several juveniles of male (10.3–17.4 mm) and female (9.9–15 mm) of *Metapenaeus* affinis were collected in trawl net catches from

Marine Bay, Andaman Island. The species is also available in west coast of India, Sri Lanka, east coast of India, Malaysia, Hong Kong, Japan and Australia.

One male (11.1 mm) and female (13.8 mm) of *Metapenaeus burkenroadi* were obtained from trawl net catches of Marine Bay, Andaman Island. The species is also found in west coast of India, Sri Lanka, Hong Kong, Japan and Australia. The distance between the rostral teeth are variable in specimens from Palk Bay, Gulf of Mannar and Andaman Sea. The pubescence on the dorsal surface of the carapace is more prominent in females than in males. Females have first four and last two abdominal segments less glabrous than those of males, although in both the sexes the fifth and sixth segments are more pubescent than the preceeding segments.

Prawn Fishery in the Krishna Estuarine Complex

The Krishna, one of the major rivers originate near Mahabaleshwar on the Western Ghats about 1000 m above the sea level, runs eastward a distance of about 2000 km across peninsular India before opening into Bay of Bengal south of Machilipatnam. Prior to its confluence with the sea the river divides into four branches.

The coast-line to the west of the western most river mouth forms a bight. The entire coast is sandy. The continental shelf is extremely narrow, being only 9.7 km. Along the coast, the irrigation canals and drains opening into the sea form mini-estuaries and there are also numerous brackish water creeks and mangrove swamps. The lower reaches of the canals and the associated mangroves form a good habitat for estuarine palaemonids and nursery grounds for a number of penaeid species. Because of the large bight west of the river mouths, the coastal waters off Nizampatnam and Kothapalem, which are relatively calm, support a number of shrimp species like *Acetes* spp.

The wide fluctuation in salinity is largely due to the dilution of coastal waters (5.35–34.65 ppt) by freshwater discharge from various canals. Although the south-west monsoon sets in around June at the upper reaches and flood water fill the lower reaches of the river soon thereafter, there is not much effect on the coastal waters until August–October. There is a fall in the salinity under the influence of the south-west as well as the north-east monsoon until a minimum salinity is observed during December-January. Thereafter, as summer approaches, there is a rise in the salinity but in April there is sharp fall due to release of reservoir water into the canals for irrigation. Maximum salinity of 32.88 ppt is observed in June.

Fishing for prawns is carried out in inland waters, back waters, coastal waters and to a limited extent in waters of the inner half of the continental shelf.

A total of 47 species belonging to 19 genera and 9 families of Decapoda, Natantia have been recorded. The individual contribution of every one of the commercially exploited species is not significant. They are:

Family–Penaeidae

Penaeus monodon, Penaeus semisulcatus, Penaeus indicus, Penaeus japonicus, Metapenaeus monoceros, M. affinis, M. brevicornis, M. dobsoni, M. lysianassa, Parapenaeopsis sculptilis, P. hardwickii, P. nana, P. uncta, P. stylifera

Family Sergestidae

Acetes indicus, A. erythraeus, A. japonicus, A. sibogae

Family–Palaemonidae

Leptocarpus potamiscus, Exopalaemon styliferus, Nematopalaemon tenuipes, Macrobrachium lamarrei, M. malcomsonii, M. equidens, M. rude, M. scabriculum, M. nobilli, M. rosenbergii

Family–Hippolytidae

Exhippolysmata ensirostris

Family–Solenoceridae

Solenocera crassicornis

The species which do not contribute to commercial fishery in the area are:

Family–Penaeidae

Penaeus merguiensis, Parapenaeopsis maxillipedo

Family–Palaemonidae

Palaemon concinnus, Palaeander semmelinkii, Macrobrachium johnsoni

Family–Pontonidae

Periclimenes demani

Family–Alpheidae

Alpheus malabaricus, A. euphrosyne

Family–Hippolytidae

Latreutes micronatus, Lysmata vittata

Family–Crangonidae

Pontophilus hendersoni

Family–Atyidae

Caridina weberi, C. gracilirostris, C. propinqua, C. bengalensis, C. rajadhari.

Ashtamudi Back Water

The Ashtamudi is a palm-shaped brackishwater system having 8 branches situated in Quilon District. With a total extent of about 32 sq. km water area, it also forms an estuary of River Kallada connected with the sea throughout the year. Salinity varied from 8.12 to 32.43 ppt.

Fifteen species of prawns were recorded from Ashtamudi back water system which included one species of sergestid, 5 species of carideans and 9 species of penaeids. The penaeid prawns are chiefly represented by juveniles of *Penaeus semisulcatus, P. indicus, P. latisulcatus, Metapenaeus dobsoni* and *M. monoceros.* While *M. dobsoni* and *P. indicus* are the dominant species in most parts of the backwaters, *P. semisulcatus* occur more abundantly in some of the deeper areas having relatively higher salinities (9.42-32.11 ppt). *P. latisulcatus* form one of the common components of the catches taken during the post monsoon period. *Parapenaeopsis stylifera*, the most dominant species contributing to the marine fishery of this area is encountered only in stray numbers.

The green tiger prawn *P. semisulcatus* is considered to be a less common species in the west coast of India, although it has flourishing fishery in some parts of the east coast. Surprisingly enough this species form nearly 40 per cent by number and rank first in terms of weight from the area. Its occurrence is also found to be more in the southern deeper areas (2 -3 m depth) where the bottom is muddy, mixed with plenty of fine black sediment. The salinity of the water is also relatively high (9.42-32.11 ppt) in this part of the backwaters when compared to the interior shallow areas. Their abundance are remarkably high during November and dwindle off by April. The size ranged from 11 mm to 85 mm, but the bulk of the catch is constituted by the size group 16-70 mm. Smaller juveniles measuring, 21-30 mm predominate during November, indicating that active recruitment of the species into this habitat might have taken place during the beginning of the North-east monsoon.

Penaeus indicus form about 11 per cent of the total collection with a size range from 12 to 95 mm with majority of prawns belonging to the size group 26–55 mm. Early juveniles of this species occur in large quantities in the near shore areas of these backwaters and predominates in the population occupying within 2 metres depth.

Fairly good number of the juveniles of *Penaeus latisulcatus* (36–60 mm) were recorded during November, but subsequently the species was totally absent from the catch. This is one of the very rare species of penaeid prawns reported from Indian waters and does not support any commercial fishery either in the marine or estuarine regions of the country.

Juveniles of *M. dobsoni* are the most abundant in the neighbouring backwaters and estuaries. In Ashtamudi back waters over 45 per cent of the catch is contributed by this species and rank first among the penaeid prawns in terms of number. Like *P. indicus*, the species is encountered at all salinity realms of the ecosystem. Its overall size range is 11-58 mm, but the bulk of the catch is contributed by the size group of 21-35 mm. In November the population are almost exclusively made up of smaller size groups indicating active recruitment of the species around this period. The early juveniles of this species below 35 mm occur in maximum abundance during August-January.

Metapenaeus monceros occur in poor numbers with a size group of 31-70 mm. Other species such as *P. canaliculatus, P. monodon, M. affinis* and *P. stylifera* are extremely rare in the region.

P. semisulcatus has not been so far recorded as a major constituent of the prawn population from any of the brackishwater environments. Juveniles of this species measuring 3.2 to 17.0 mm in carapace length live in areas of the sea where *Zostera marina* are growing. The large concentration of juveniles of this species and the occurrence of the king prawn, *P. latisulcatus* in appreciable numbers in this back water system would suggest that it is an ideal nursery ground for these species. Abundant resource of the young ones of *P. semisulcatus* also shows the possibility of the existence of a good breeding stock in the adjoining sea where the adult prawns have started appearing in the commercial catches in minor quantities. *P. stylifera* the most dominant species contributing to the marine fishery of this area is encountered only in stray numbers in the back waters in confirmation with the habit of the species.

Backwater Fishery

Metapenaeus dobsoni ranks high among the commercially important species of penaeid prawns occurring along the south-west coast of India, contributing a good portion of the catches practically throughout the year. It is one of the two species making up the bulk of the catches from the back waters and canals (brackish water) stretching along the southern half of this coast, wherein prawn fishing is carried out on all the year round. In the northern regions of Travancore-Cochin, the paddy fields bordering the backwaters serve as rich prawn fishing grounds from about the middle of November to the middle of April every year, when no paddy cultivation is done.

M. dobsoni is the dominant species in the prawn fauna of the backwaters in all but three months. The major portion of the catches is obtained by means of stake-nets, Chinese nets and cast nets. Stake-net fishing is carried out, wherever the tidal flow is strong, the nets are usually fixed at the commencement of the ebb-tide and hauled up when high tide sets in. Though type of fishing is not exclusively meant for catching prawns, they nevertheless form the bulk of the catches. The fishing continues throughout the year except when there are high floods during the monsoon period, June, July and August.

Like stake nets, Chinese net also work throughout the year in Cochin backwaters. Cast nets are used in the shallow regions of the backwaters and in the network of canals connected with them. Prawns form the major portion of the catches. During monsoon months, however, canal water becomes practically fresh and prawns are quite scarce. Prawns are also caught in drag nets and by hands.

Marine Fishery

The marine prawn fishery is largely seasonal. In June and July and again in November and in the two following months *M. dobsoni* is relatively abundant in the commercial catches. Prawns and a variety of small fish are caught in boat seines during the warmer months of the year, but the former are seldom obtained in large numbers. But they are found in considerable numbers in inshore waters (2-4 fathoms) concentrations, discovered near the shore. Cast net may be used if the prawns occur quite close to the shore.

At Narakkal, the fishery starts towards the close of April or early in May and continues till about the end of September. After the first week of October, however, it declined abruptly and on several days thereafter, hardly any prawns have been caught, though the boat seine had been in use on most days till about the middle of January. The catches of prawn at this place are seldom very heavy.

The rich colonisation (717 juveniles per 100 sq metre) of early juveniles of *Penaeus monodon* were noticed over the sea grass beds in Kovalam back water, near Chennai. This congregation could be attributed to the clinging habit of the species in its earlier part of life. This habit of taking shelter among the grasses may help them to escape from predators. *P. monodon*, also known as the grass shrimp takes refuge among the weedy areas in estuaries during its post larval and early juvenile stages. Taking advantages of the clinging habit of fry of tiger prawn, successful fry collection have been achieved by suspending bunches of twigs and grass from a horizontal rope in the estuaries and bays.

Penaeid prawns of commercial importance from Porto Novo waters are *Penaeus indicus, P. monodon, P. semisulcatus, P. merguiensis, Metapenaeus dobsoni, M. monoceros, M. brevicornis* and *M. affinis*.

All the species of prawns are not using the brackish water environments as their nursery grounds. Considerable variations occur among the prawns both in the degree to which the brackish water environment is put to use by each species in its life history and, in the distribution of the parents and juveniles population in the marine brackish environments.

The estuarine, back water and mangrove environments play a vital role in the life cycle of prawns and in the establishment of its fishery.

In Porto Novo aquatic biotopes, such as, estuary, backwater and mangrove, are very rich in prawn fauna. They provide a very good recruitment to marine fishery.

Penaeus indicus is very common in all the biotopes. It showed the primary peak of abundance from January to May and a secondary peak in September to November in the estuary, backwater and mangrove. More number of juveniles were collected from the backwater than from the other biotopes. Large number of post larvae of 8-15 mm sizes were collected from the shallow pools near the mouth of the river. *P. indicus* showed the maximum percentage composition in the mangrove region. The maximum size of this prawn caught from the estuary was 142 mm (total length).

P. semisulcatus was very common in the estuarine and backwater regions and was scarce in the mangrove areas. They were usually found abundantly in the marine grass beds of *Cymadocea isoctifolia*, than in other area. They occur from May to October in estuarine and backwater environments. The peak was during August and September. The maximum size observed was 102 mm.

Penaeus merguiensis, commonly known as banana prawn occurs throughout the Indo-Pacific in tropical and sub-tropical waters supporting commercial fisheries along the coasts of Malaysia, Thailand, Indonesia, Australia, India and Pakistan. The species is one of the commercially important penaeid prawns occurring in Indian waters. It inhabits the coastal waters upto a depth of about 55 metres. It is relatively more abundant in shallow areas where the sea bottom is muddy and sandy. On the west coast of India this species is caught in appreciable quantities from North Kanara, Goa and Ratnagiri coasts. It forms 1.4 per cent of the total prawn landings by shrimp trawlers at Karwar. A seasonal fishery for this species from June to August has been observed in Goa. Fishermen of North Kanara coast operate gill nets and shore seines in coastal waters throughout the year at different depths to catch this species depending on weather conditions. The occurrence of banana prawns in good quantities have been noticed in the bottom set gill nets operated in the inshore waters off Sankrubag landing centres.

The bottom-set gill nets are employed for catching prawns along the north Kanara coast during the monsoon period, June to August as in other regions of the west coast. The prawn grounds covered by the fishery extends upto about 2 km from the shore where depths ranges from 6 to 9 m.

The fishery season generally commences with the onset of monsoon and extends for about three months in June, July and August. Peak of the fishery varies from year to year within this period.

Taking into consideration the entire fishery of this coast, maximum landing of *P. merguiensis* was recorded at Ambekodar during 1984 while the best catch per unit effort was observed at Sankrubag during June-July, 1982.

The total length of this species varied between 111 and 165 mm for males, and 106 and 215 mm for females. The size frequency was generally unimodal or biomodal in nature. The bulk of the fishery was supported by the size groups between 130 and 170 mm. The predominance of females over males were noticed in all the centres with an exception at Chendia during 1981-82 when males out numbered the females. Most of the females observed in the catches did not have mature ovaries.

Some kind of sporadicity of occurrence and schooling behaviour of the banana prawns have been observed along the inshore waters of Karnataka coasts during monsoon period, June-August *P. merguiensis* is capable of undertaking long range migrations moving even upto 150 km off shore.

P. merguiensis migrate in shoals from south to north along the coast. Juveniles of *P. merguiensis* show migrations from back waters to the coastal waters in Karwar region during January-May. It is possible that juveniles migrating from backwaters during this period remain in the coastal waters until they attain the adolescent stage and then move out to deeper waters.

The mechanized fishing industry in South Kanara coast has gained a great momentum, from a mere 20 trawlers during 1962-63 to about 800 in 1971-72. They were of 6.6 to 13.0 m in length fitted with 20-85 HP engines and using otter trawls of 9-21 m head rope length and 2.5 cm stretched mesh at the cod end in depth of 10-20 m. The trawling season lasts from the mid October to May.

The annual average catch of prawn and fish was seen to be 952.5 and 1755.6 tonnes respectively. Prawns formed about 35 per cent of the catch. The catch rate of prawns declined from 138.2 kg during 1962-63 to 35.5 kg during 1970-71. The maximum catch rate for prawns observed in different quarters of the years varied from 52.8 kg (1969-70) to 219.9 kg (1962-63) at Mangalore and from 13.5 kg (1969-70) to 102.6 kg (1963-64) at Malpe.

Metapenaeus dobsoni always dominated the prawn catches (52.3 per cent) followed by *Parapenaeopsis stylifera* (31.8 per cent) and *M. affinis* (11.6 per cent). Other prawns *(Penaeus indicus, P. monodon, Acetes* sp) showed considerable annual fluctuations.

The modal sizes of males and females of *M. dobsoni* ranged from 68 to 103 and 68 to 118 mm respectively. They constituted late '0' to 1–year group. Recruitment of smaller sizes was found to take place during December–February and a progressive increase in the modes upto 93 mm for males and 108 mm for females was noticeable till May. In the cast net catches of August-September a further increase in the modal size to 103 and 118 mm for males and females respectively was observed. Such bigger size groups sometimes continued to occur in the trawl catches till January. In the case of *P. stylifera* the modal sizes exploited by the trawlers were more or less similar and the recruitment takes place twice during the season, *i.e.* November-December and April-May as indicated by the occurrence of smaller sizes in those months. The species did not occur during August-September. As regards *M. affinis* the modal sizes exploited ranged from 83 to 148 mm for males and from 88 to 163 mm for females, that is, from '0' to 2-year groups. Recruitment of smaller sizes was noticed during March-May.

The monsoon prawn fishery along Mangalore coast largely depend on the weather conditions as well as on the availability of shoals. Prawns alone formed 63.5 per cent catch in miniature purse seine which are generally confined to near shore waters within 15 m depth. Among the species, *Metapenaeus dobsoni* contribute 96.8 per cent of the prawn landings, while *Penaeus indicus* and *Parapenaeopsis stylifera* together form the rest. 97.9 per cent prawn was landed at Ullal in August.

Around Panambur harbour, prawns formed only 22.1 per cent of the catch of this centre. *M. dobsoni* is the principal species contributing 95.0 per cent of the prawn landings and *P. indicus* form the rest.

In other landing centres like Uchila, Hejamadi and Malpe *M. dobsoni* contributes 100 per cent catch; while in Polippu, *M. dobsoni* contribute 92.1 per cent of the population and rest by *P. indicus*.

The fishery in Mangalore coast is supported by large sized prawns. *M. dobsoni* is represented by sizes ranging from 68 to 98 mm in males and from 58 to 118 mm in females in Ullal area. At Baikampady prawns ranging in size from 73 to 98 mm in males and 78 to 113 mm in females supported the fishery.

Among males 0-year class formed only 3-4 per cent at Ullal and about 2 per cent at Baikampady, whereas, 1-year group (above 80 mm size) contributed the bulk of the catch (96-97 per cent at Ullal and 98 per cent at Baikampady). Among females, 22-33 per cent was in 0-year class at Ullal and 22 per cent at Bailampady, while 1-year class (above 95 mm size) formed 66-78 per cent at the former centre and 78 per cent at the latter.

The overall sex ratio in *M. dobsoni* indicate that males outnumbered females at both centres. In July males forms 61.5 per cent and 54.1 per cent at Ullal and Baikampady respectively. However in August 86 males and females were distributed more or less equally.

During 1986 season about 32 per cent of females were in mature condition both at Ullal and Baikampady. Impregnated females formed 17.7 per cent and 14.3 per cent respectively at these centres. In August, 18.2 per cent of females were in impregnated conditions at Ullal. The occurrence of spent females at Ullal (6.5 per cent in July and 16.4 per cent in August, 86) suggested peak spawning particularly in latter month.

Exceptionally heavy catches of *M. dobsoni* were obtained in miniature purse seine (Matabala) on certain days in July and August, 1986. The species alone contributed upto 98 per cent or even 100 per cent of the prawn landings. It was estimated that around 242 tonnes of prawns were landed between Ullal and Malpe, within a range of 70 km in a short period. On the question of adverse effect on the resource due to heavy catching of these prawns at the time of peak spawning period, it can be seen that monsoon fishery is exclusively supported by large sized prawns with modal lengths at 88 mm in males and 103/108 mm in females. As most of these prawns have already spawned 2-3 times and also reached their maximum size, it is desirable to catch them during that period, instead of leaving them to breed again. Moreover, this prawns with such a short life span (maximum of 2 years) are available to the fishery for not more than a year or so. Hence it is possible that they may die of natural mortality if not caught at that size, which may be a loss to the fishery.

Although the catches were heavy on certain days in July, 1986, the total landings of *M. dobsoni* was only 106 tonnes at Ullal and less than 250 tonnes along the entire Mangalore coast.

Metapenaeus dobsoni constitutes more than 50 per cent of the prawn catch by the trawlers at Mangalore, Karnataka coast. It forms an appreciable fishery in the indigenous catches also, especially during the south-west monsoon. Juveniles of this species are caught in considerable numbers in the estuaries during August-February. The species appear in the coastal belt up to 22 m. depth.

The fishery is dominated by larger size groups (above 95 mm) during July–November. Recruitment of smaller size groups (60 mm onwards) is observed during December-January. The seasonal size oriented movement of the species in the fishing grounds have been observed in south-west coast of India.

M. dobsoni breeds throughout the year with peak spawning periods during April, June and November-December. Several broods enter into the fishing grounds and in a season, a maximum of five broods could be noticed to appear in the fishing grounds.

The pattern of growth is not the same during the different years, which can be attributed to the variable ecological conditions. Males and females of *M. dobsoni* grow to a length of 85 and 105 mm and 95 and 120 mm during the first and second year of their life respectively. During the third year very little growth seems to take place since the species attains the maximum size by then. In Kerala coast, the males and females of this species respectively attain a length of 97 and 115 mm and 122 and 138 mm at the end of first and second year of life. The estimated size of *M. dobsoni* for all sexes in Kerala coast is 95, 114 and 118 mm at the end of one, two and three years respectively.

The fishery of *M. dobsoni* at Karnataka coast is primarily constituted by the late 0-year class (above 60 mm) and one year old prawns.

The females predominates the population. There is frequent incidence of females exceeding males by more than 1.5 times during April-May and November-December, which coincides the peak periods of breeding. Significant variations in the sex distribution is found to be associated with segregated movements for the breeding purpose in Kerala coast.

Matured females are present in almost all the months indicating that this species breeds throughout the year. Peaks of spawning activity are evident in January, April and October-November. The smallest mature female was 71 mm. Impregnated specimens are encountered in most of the months, the smallest one measuring 68 mm. The percentage of impregnation is found to be generally high in May, August and October.

The coast line of Kutch, between 22°45' to 23°50' N and 68°24' to 70°13' E stretching about 320 km harbour a number of prawn species and form good source of prawn fishery. The sea water enter in Kandla creek and spread out 8 km behind the Little Rann of Kutch during high tide offer a good habitat for prawn migration.

Higher salinity (35.9-55.8 ppt) was recored throughout the year except September and October (2.7–7.8 ppt) during which salinity showed an inverse relation with prawn production. During winter high production of prawn was supported by low salinity.

There are two fishing seasons for prawn in Kutch region. Monsoon fishery is conducted from August to October, but starting depend upon the onset of monsoon. Winter fishery starts from October onwards upto the end of February. At winter season big sized prawns are caught from the mud flats.

Metapenaeus kutchensis, M. brevicornis, Penaeus indicus, Parapenaeopsis stylifera and *Paleamon stylifera* are the major catch constituents in Kutch region. The bulk of winter fishery in the Gulf areas is consisted of *M. brevicornis* (60 per cent), *M. kutchensis* (20 per cent) *P. indicus* (10 per cent), *Parapenaeopsis stylifera* (5 per cent) and *Paleamon stylifera* (5 per cent). The number of *M. kutchensis* decrease from August to October due to increase in size. Influx of juveniles (15-75 mm) of *M. kutchensis* have been observed during December to April and June-July with 26-46 per cent males and 54-74 per cent females. The larger groups of 95-151 mm assembled further interior of the Gulf during December to April had 25–40 per cent males and 60-75 per cent females in the population.

Prawns constitute the major portion of the total marine fish catch of the Gulf of Khambhat, between long 72°2' to 72°6' E and lat. 21° to 22°2' N along Gujrat coast. The average tidal fluctuations in the Gulf (open sea) varied from 2.56 m (neep) to 9.36 m (high) and the tidal range on the coasts varied from 1.3 m to 5.20 m. *P. sculptilis* and *M. dobsoni* are permanent inhabitants of the Gulf of Khambhat which form a good nursery, with estuarine areas on both the coastal belts.

Parapenaeopsis sculptilis constituted 47.3 per cent and *Metapenaeus dobsoni* 51.6 per cent on an average throughout the year. *P. sculptilis* was relatively abundant in November and December 1978, January, June–September 1979 and May–July in 1980. *M. dobsoni* was abundant in August 1978, February to May, 1979 and October to February, 1980. The largest of *P. sculptilis* encountered was 147 mm and that of *M. dobsoni* was 124 mm. The juveniles of both the species measuring 20 to 40 mm were found in most of the months.

The abundance of *P. sculptilis* was found to be directly proportionate to the dissolved oxygen concentration, while no remarkable variations were found for *M. dobsoni* even at different oxygen concentration. This suggests that *P. sculptilis* is sensitive and *M. dobsoni* is tolerant to low oxygen concentration.

Five maturity stages were distinguished in the ovarian development of *P. sculptilis*. These are (i) opaque and slender, (ii) light green and slightly elongated, (iii) dark green and distended, (iv) dark green granular and fully distended and (v) orange-yellow and flaccid. In *M. dobsoni* four stages namely, (i) opaque and slender, (ii) dirty-white and slightly distended, (iii) pigmented and fully distended and (iv) whitish yellow with less pigments were discernable. Some females of both the species had a noticeable expansion of the posterior lobes of the ovary in the region of the first to third abdominal segments.

The females of *P. sculptilis* attains the ovarian development at the size of 70 mm and that of *M. dobsoni* at 55 mm. Majority of the females of *P. sculptilis* with developed ovaries were collected during August, September, October and March-April indicating the spawning season during these months. *M. dobsoni* females measuring 55 mm to 120 mm spawn during April, May and June. The presence of *M. dobsoni* juveniles measuring 20-40 mm in most of the months is suggestive of prolonged spawning of *M. dobsoni*. Females of *P. sculptilis* spawn in the inshore waters.

The prawn fishery in Maharashtra, unlike in other maritime states in India is supported by a number of species, each contributing to a fishery of varying magnitude.

The prawn fishery by indigenous 'dol' nets generally commences in the area by September-October, after the south-west monsoon and ceases by late May or early June at Versova, whereas it continues unabated at Sassoon-Docks. These nets are operated at a depth of 30-40 m at Versova and at 10-15 m at Sassoon Docks.

The best catches and highest catch per unit effort were obtained in 1970-71 at Versova (2330 tonnes and 286.6 kg respectively), where as at Sassoon Docks it was in 1967-68 (1917 tonnes and 114.2 kg respectively). As against a mean yearly landing of 3173 tonnes and 2153 tonnes respectively at these centres for the period 1959-63, a decline in the yield by over 30 per cent at Versova and 50 per cent at Sassoon Docks is noticeable over the years (1325.2 tonnes at the former centre in 1969-70 to 1975-76 and 1462.2 tonnes at the latter centre in 1966-67 to 1971-72).

The peak catches were obtained mostly during October-December and March-May at both the centres.

The species contributing to the fishery include, *Metapenaeus affinis, M. brevicornis, Parapenaeopsis stylifera, P. harwickii, P. sculptilis, Solenocera crassicornis Atypopenaeus stenodactylus* and non-penaeid prawns like *Acetes indicus, Palaemon tenuipes* and *Hippolysmata ensirostris*. The non-penaeids contributed to the bulk of the prawn catch and formed about 70-80 per cent of the prawns caught in dol nets. *Acetes indicus* and *Palaemon tenuipes* formed the major constituents of the non-penaeid catch and one or the other was found to dominate in the fishery during the different months. The sergestid shrimp, *Acetes indicus* was by and large the most important species that contributed to the bulk of the annual catch in all the seasons at both the centres except during 1969-70 when *P. stylifera* was landed in fairly large quantities.

During October-December and March the landing of the dominant species *Acetes indicus* is considerably high at Versova. At Sassoon Docks, the catches of this species is generally high during November–December. Bulk of the *Palaemon tennipes* is landed during March-May when more than half of the annual catch is recorded at both the centres.

The catches of *Metapenaeus affinis* is generally better during September-November and April-May. October-December appears to be the peak season of *P. stylifera*. Catches of *P. hardwickii* are relatively high during November-December and in March at Versova. Peak catches of *Solenocera crassicornis* are generally obtained during October-November and March-April at Versova.

Penaeid prawns like *M. monoceros, Penaeus monodon, P. merguiensis, P. penicillatus, P. japonicus* and *Metapenaeopsis stridulens* were found to occur in small numbers in the catches at Versova, where dol nets are being operated at slightly deeper waters. The landings of *M. monoceros* and *P. merguiensis* (size ranging from 150-180 mm both in cases) were usually high on certain days in November and December, 1975, *Metapenaeopsis stridulens* was landed in sizeable quantities in November, 1971 at Sassoon Docks.

In addition to the above mentioned species of prawns, penaeids like *P. semisulcatus, P. canaliculatus, M dobsoni, Metapenaeopsis mongiensis* and non-penaeids such as *Palaemon styliferus, Hippolysmata vittata, Macrobrachium rosenbergii, Macrobrachium idae* and *Tozeuma* sp. were recorded in stray numbers.

The penaeid prawn, *Parapenaeopsis hardwickii* forms a fishery of commercial magnitude along the coast of Maharashtra and to a lesser extent in Andhra coast. The species is caught in stake net (Dol net) operating at 30-40 m depth at Versova and in 'dol' and trawl nets operating at 10 -15 m and 30–40 m depth respectively.

The fishery of *P. hardwickii* commences usually in late September or early October. It formed 3.9 per cent and 0.6 per cent of the catch of Versova and Sassoon Docks respectively. The heavy catches during November and December decreases thereafter.

Males grow very slow whereas females register faster rate of growth (monthly growth of fernale is 7.6 mm in Sassoon Docks and 5.4 mm at Versova)

In the female of *P. hardwickii,* five maturity stages were recognized as immature with transparent ovary, early maturing with yellowish ovary, late maturing with orange coloured ovary, mature with deep orange coloured ovary and spent recovering with dirty yellow or white ovary.

The spawning season of *P. hardwickii* appears to be prolonged with peaks during October and April-May at Versova and during November–December at Sassoon Docks. The spawning season of the species is protracted from October to February with peaks in December and January in Bombay waters.

Though spawners are available above the size of 63 mm in length at Versova and 68 mm in length at Sassoon Docks the bulk of the spawning population was in the size range of 78-108 mm and 78-113 mm respectively at these centres of 0-year and one-year olds. 0-year olds are the main spawners during November-December and April-July, while one year olds during rest of the period at Sassoon Docks. Similarly 0-year olds are the principal spawners during November-February and April, whereas the bulk of the spawners belong to one-year old spawns during March, May and October at Versova.

The migratory habit of the prawn for spawning to slightly deeper waters at 30-40 m depth off Versova was noticed.

Though the smallest female with fully matured ovary measure 63 mm in length, the majority of them are found to be mature at 73 mm and above.

There is preponderance of females over males. During certain months males were practically not observed in the catches and the trawl catches were solely comprised of females.

The fecundity in sizes ranging between 63 mm and 121 mm varied from 17250 to 1,21,570 eggs with an average of 63,690 eggs.

Chapter 7
Krill, Spotted and Oppossum Shrimps

Dense schools of *Euphausia superba* mill at the surface close to the icy Antarctic shore, swarms with the reddish, thumb-length crustaceans that compose the predominant krill species of south polar seas.

Each cubic metre of the ocean holds tens of thousands of krill, all of a size and oriented like marching soldiers, surging forward in their disciplined legions.

In defense against human intrusion, they instantaneously moulted before they fled, leaving behind as decoys their empty husks, a ghost school in an empty sea. Congregating in millions, krill can stain the seas as if with blood.

The name "Krill" derives from *kril* an old Norwegian word once applied to tiny creepy–crawly things, lively vermin, and larval fish. To day krill means whale food, and a number of species of euphausid shrimp, as well as other very small planktonic animals, feed the biggest creatures of all, the great baleen whales. Krill also are harvested by man and processed into feed for livestocks, poultry and farmed fish. Their swarms constitute the ocean's richest source, as yet barely tapped–of protein.

E. superba is the almost exclusive food of the southern baleen whales, some species of which are now pushed to the edge of extinction. To day, because of the depletion of Antarctic whales, the shrimps that nourished them are much more numerous than 50 years ago. Marine biologists calculate that the potential annual yield of this "unutilized whale food" could exceed the present world harvest of all other edible marine species combined. It is estimated that 50 to 150 million tonnes of Krill conceivably could be taken by humans each year.

Although in Antarctic waters the term "Krill" refers specifically to *E. superba* shrimp, in different areas of the world's oceans it designates a variety of animals, depending upon what a particular species of whale consumes in certain region. Off Vancouver Island, Krill means the vast shoals of mysid shrimp, the prey of gray whales. In the Chilean fjords, a thumbnail-size pelagic red crab of the genus *Munida*, the lobster krill, forms immense swarms and is a favourite food of sei whales. In the North Atlantic and North Pacific Oceans, krill includes schooling small fish. But in Antarctic waters *E. superba* so abounds that the baleen whales feed on it almost exclusively.

All the planktonic animals called krill congregate in enormous schools, primarily in polar and subpolar seas. Schools of krill, widely scattered by other predators, such as, fish and seals, are often simply hard to locate.

E. superba shrimp have eleven pairs of legs. They swim with five posterior pairs, which are broadly paddle–shaped, and they feed with the six forward pairs. Each feeding leg, split into two branches, carries stiff bristles and feathery setae. The darting shrimp move their legs so fast to see a blur.

Euphausid shrimp are essentially herbivores, feeding on diatoms, the tiny, single-celled plants or phytoplankton, that float in greater abundance in polar seas. In the sea, the krill schools use their highly developed sensory receptors to find food. Encountering an enticing taste or smell, they feed by repeatedly throwing wide their legs to enfold a packet of sea water that smells edible and presumbly contains food. The krill then squirt the sea water side ways through their setal filters, entrapping algae in a feeding pattern much like that of baleen whales.

Millions upon millions of shrimp sweeping past in response to the secret signals that comand the unison of schooling creatures, flowing uninterrupted like a band of army ants hugging the under-surface of the sea. When startled, the shrimp fled back-ward at blinding speed, with powerful flicks of their tails. Each individual swam rapidly, as if lost, until it found another isolated shrimp. Then they swam as a pair. The two became four, and then more and more came together, as the scattered shrimp regrouped into the cohesive security of the school.

The life cycle of *E. superba* is attuned to the seasons in Antarctic seas. These shrimp mate in austral spring, October and November. As with most crustaceans, because of their rigid exoskeletons the male must wait to implant his sperm until a female has moulted her old shell.

A female of *E. superba* produces many sets of thousands of fertile eggs throughout the summer spawning season. Released into the sea, these eggs sink hundreds of metres to depths where few predators dwell. There the eggs hatch and the larvae, or nauplii, looking nothing at all like the adults, develop in relative safety. Eventually, however, their yolk-sac reserves are depleted and the nauplii must swim towards the surface to find the phytoplankton on which they will feed.

E. superba develops slowly in the icy Antarctic waters, passing through many moults and five life stages from nauplius to adult. Growth to a mature length of about six centimeters may take 3 to 4 years. All this time the Krill stay together in enormous schools, cruising great distances at sea and beneath the ice in search of food while avoiding their many enemies.

Antarctica's *E. superba* represents a potential source of food for humans. Dried Krill are more than half protein and rich in vitamins especially vitamin A. Japan and Soviet Union already harvest heavily in south polar waters. In the austral summer of 1981-82 they took as much as 500000 tonnes with the USSR catching the bulk, principally for the dry meal to feed domestic livestock and poultry.

Japan conducts two Krill fisheries, one in Antarctic waters for *E. superba* and the other along the coasts of the home island of Honshu for the smaller, *E. pacifica*.

Fresh, uncooked euphausids have almost no taste. Frozen or dried Krill develop a strong, rather discouraging flavour.

In Japan frozen euphausids are used to feed trout, salmon, sea bream, red-snapper and yellow tail in extensive fish farming operations. Vitamin A, carotenoids in the shrimp enhance the tone and pigment of fish flesh. Sport fishermen also buy frozen Krill to use as chum, most fish find euphausids delectable.

In north coast of Japan, compact schools of *E. pacifica* rise to the surface close to the coast. They move quickly when disturbed.

At sunrise the fishermen watch for the reddening of the water and when a school is sighted, the boats close in, recklessly jockeying for position. The fore deck of each vessel is bare except for two giant poles set forward across the bow bulwarks, with a surprisingly small net slung between them. The two poles slide forward like probing antennae. With the captains order the booms tilt and plunge, and the net spreads open beneath the bow. The vessel pushes the net slowly through the school and engulfs the Krill. The cod end, the collecting bag of the net, fills, and the crew winches it aboard.

This simple technology and the small boats used to catch *E. pacifica* contrast sharply with the modern Antarctic *E. superba* fishery. Antarctic schools are huge, and often swarm at depths as great as 150 to 200 metres. The bigger southern Krill, *E. superba* swim faster than other species and readily avoid small nets. Fishing vessels are designed to counter these capabilities.

Big stern trawlers towing enormous mid-water trawl as wide as 80 metres and 50 metres deep at the mouth. Otter boards, wooden slabs, the size of barn doors slip side ways through the water away from the ship to spread wide the mouth of the net, which sweeps an area almost equal to a foot-ball field on edge pulled through the sea. A haul of 8 to 12 tonnes of Krill is common.

Early morning in January, three Soviet trawlers were found pulling their immense nets along parallel tracks through a massive patch of Krill at some hundred metres depth. The cod end was so heavy and swollen with Krill dwarfing the Soviet seamen as it inched up the stern ramp. Crewmen hooked the end of the net to an overhead hoist, and the Krill poured into the hold.

The Russians operate about a hundred and the Japanese 14 of these giant trawlers in the Antarctic Krill fishery.

Excessive Krill fishing could be disastrous for the entire fragile Antarctic ecosystem, because almost every animal there depends for survival directly or indirectly on *E. superba.*

Perhaps overfishing of Krill will not occur in the Antarctic waters. Costs and distance may curtail the harvest. The Antarctic waters are remote from the countries best able to exploit their resources, so it is difficult to operate a cost-effective fishery there. Only the Soviet Union is presently engaged in large–scale fishing for Antarctic Krill. Several other nations that raced each other to exploit Krill are now restricting their efforts.

Estimates vary from 125 million tonnes to 6 billion tonnes of Krill the Antarctic seas can support. The fishing season also coincides the Krill breeding season. Much more knowledge of overwhelmingly important Krill species are required besides international cooperation, patience and good will to protect the bountiful, elegant, clever little shrimp, *Euphausia superba*

Billions of animals, such as Krill and plankton constantly plough through the ocean's waters from the tropics to the poles. Scientists reported in 2006 that large schools of Krill stir a well stratified fjord along the coast of British Columbia, Canada. Another study, published in the same year calculated overall amount of energy supplied by the swimming motion of the Krill was substantial.

Acetes indicus, a sergestid shrimp is estimated to contribute about 20 per cent of the marine prawn landings along the Maharashtra coast constitutes a seasonal fishery along with three other species namely, *A. johni, A. sibogae* and *A. japonicus.* They are mostly caught by bag net (Dol net) upto a depth of 40 m, scoop net at 3-5 m and in trawlers having very small meshed cod end.

Prior to start of monsoon and just after monsoon months, the landings of *Acetes* show an increasing trend.

A. sibogae is the dominant species in the estuarine environment, while *A. johni* indicate its preference for a marine habitat.

A. indicus measuring 13-40 mm, *A. johni* of 15-28 mm, *A. japonicus* of 12-25 mm and *A. sibogae* of 9-33 mm form the commercial catches in the fishery.

In all the four species, common items of food consists of copepods and appendages of decapod crustaceans (60 per cent), foraminiferan and molluscan shells and shell fragments (10 per cent) sand grains (10 per cent) and debris (20 per cent).

Maturing females of *A. indicus* were noticed in January and of *A. johni* in March. Females of *A. sibogae* with maturing ovaries were observed in April-May indicating possible breeding during monsoon months. In all the three species a minimum size of 13 mm was recorded for females with ripening ovary.

In Indian coast the average catch of *Acetes* is estimated at 14500 tonnes constituting 11.2 per cent of total world *Acetes* production.

The average life span of Acetes is less than six months and adult dies soon after spawning. In India a small portion of the catch is consumed in fresh condition and the rest is mostly sun-dried.

The Periscope shrimp, *Atypopenaeus stenodactylus* constitutes a fishery at Versova along Maharashtra coast. The fishery is seasonal lasting from November to May constituting about 3.3 per cent of total 'dol' net landings at this centre.

The area of operation for *A. stenodactylus* at Sasson Dock is off Marud-Harnai coasts of Maharashtra at a depth range of 27 to 70 m. At Sasson Dock the landings of the species by the trawlers constituted 1.5 per cent of total prawn landings.

Variations were noticed with regard to size range and sex ratio at each of these centres. Total length was from 22-38 mm at Versova with a sex ratio of 1 : 3 with female dominance. In trawler landings at Sasson Dock, females were always dominant. A size range of 40-65 mm was noticed in the specimens from trawler catches at Sasson Dock.

The dominant mode 25-35 mm was noticed in November, December and January at Versova. During other months the main mode remained at 28 mm during 1983 and 33 mm in 1984.

In trawler catch at Sasson Dock the mode at 56 had dominated in March which more or less continued upto April. It shifted to 60 mm in May.

Crustacean appendages (Copepods and Decapods mainly), sand grains and debris constituted the major food items of the species.

In trawler landings which were mostly females, the percentage of mature specimens were as high as 80 per cent. The species spawn during March-May. The sudden appearance of this species in

November in Versova suggests their shoreward migration from deeper waters where spawning takes place.

Mysids or "Oppossum shrimp" have been reported to form a fishery at Satpati. Maharashtra. The species *Mesopodopsis orientalis* is caught in stationary bag nets operated in the creek at Satpati in about 3-5 m depth. The length of the net is approximately 6-7 m and the mouth is about 2.5 to 3 m. The net is made out of fine meshed nylon cloth with a mesh size of 0.1 mm. The species are caught in April-May by 1360 to 1500 fishing units and the catch ranged from 14000 to 15005 kgs.

Along with fish larvae and larvae of decapod crustaceans, the mysid, *M. orientalis* constituted 90 per cent of the landings between 1984-1986.

The chief characters that held in the identification of the species *M. orientalis* are the structure and the shape of telson and the fourth pleopod of males. The lateral spines on telson stated to be four for this species was found to vary from 3-5. But the majority had only four spines on each side of the telson.

The fishery for *M. orientalis* is seasonal. It starts from April and closes just after the first rain in June. The size range of the species is very small varying from 5-7 mm and sex ratio was observed to be 1 : 2 with females in domination. *M. orientalis* are abundantly available during April and May.

Spotted shrimp (*Sergestes lucens*), a species of Sergestidae family and considering its size and movement, it would fall more into the category of a micronekton rather than an animal plankton. Because this species has red pigments and light-emitting tissues scattered throughout its almost transparent body, from a distance a school of these shrimp appears pinkish in colour. Aside from their appeal as zoological specimens, these shrimps are considered a delicacy for their flavour and pleasing appearance.

Spotted shrimp are purely embaymental resource. Spawning takes place from May to November with the peak in July and August. Although eggs and larvae have been found in areas throughout the bay, an overwhelming majority of spawning schools are found in a relatively small area in the vicinity of the mouth of the river.

After the larvae have hatched they are gradually spread out into the bay along with the movement of the surface water which flows into the bay from the open sea. Although a part of the larval population is carried out of the bay, the majority grow up within the bay area, and as they approach maturity and their swimming ability improves, they begin to congregate around the shores of the inner part of the bay.

The life span of these shrimp is 15 to 16 months. Within about a month of hatching and going through the nauplii, elaphocaris and acanthosoma stages, they reach the post larvae IV stage with a body length approximately 7.5 mm. Then while repeated ecdysis at a rate of once in every 3 or 4 days, within 3 to 4 months they reach a size of 20 mm and by 10 to 12 months they reach a size of 40 mm at which time they begin spawning.

During their life cycle, spotted shrimp mature while migrating horizontally from the coast to the offshore and then to the coast again. At the same time there is a daily perpendicular migration. Spotted shrimp cannot become the target for boat seining until they reach a body size of 20 mm, because not only is it when they reach this stage, they begin to gather in coastal areas in high concentration, but also they became more adaptable to environmental changes such as, water temperature and salinity, thus allowing them to increase their range of perpendicular movement.

In the daytime these shrimp form schools of several meters in diameter and descend to a depth of approximately 150 metres and at night they will ascend while diffusing the density of their schools slightly. The spotted shrimp which in this way ascend to the upper water layer at night are the ones which will be caught by tow net. The fishing ground for these shrimps form mainly in the vicinity of river mouth in an area that has an isobath at a depth of about 200 meters.

In the autumn (October-December) season, the shrimp are born, the catching ratio is about 30 per cent, in the following spring season about 50 per cent and in the next autumn season about 20 per cent. In other words within a year and half of the time they are born, almost all the shrimp population has been caught.

Chapter 8
Catching of Prawns

Catching of prawns from sea requires craft and gear. Various kinds of traditional crafts and gear, locally made, are in use since centuries for capturing different kinds of fish including prawns.

The most primitive wooden craft used for fishing in India is the Catamaran, made up of number of logs tied together. The front end is cut in a slightly slanting manner for easy cutting against the current. Besides one stout central log in the shape of a keel, there are two side logs kept raised forming a depression in the centre. The size in Gulf of Manner is 6.7 × 0.9 × 0.75 m. Catamarans are propelled by paddling with oars. Sails are also used on the craft. Though primitive in appearance, the catamaran can be dragged ashore after use without difficulty for cleaning and repairing. It will not sink in rough weather, even if the waves are high and passing over the craft.

Open wooden boats canoes are used for fishing in estuaries, backwaters and rivers. The big plank–built boats of different sizes are used in different places. The Masula boats have no ribs or frames, the planks being sewn together with coir ropes.

Motorization of country crafts for fishing started at Maharashtra in 1950's. Since, 1953 many small and medium–sized harbour boats have been introduced into the fishing industry. Gill net operations and trawling especially for prawns started on modern lines.

Introduction of synthetic fibre twines is an improvement in fishing together with introduction of larger boats with greater efficiency in the towing speed and the speed with which the gear is hauled in. The non-mechanized wooden boats used to catch prawn from coastal areas of 5 to 20 m deep, while modern motorized boats are able to exploit regions upto 400 metres in depth.

Table 11: Fishable Areas in Indian Ocean

Sl.No.	Area	Square Kilometres	
		Upto 50 m Depth	Upto 200 m. Depth
A.	**Arabian and Laccadive Seas**		
(i)	Bangladesh	9819	54953
(ii)	Gujarat	64810	99373
(iii)	Maharashtra	25512	104758
(iv)	Goa	2849	9984
(v)	Karnataka	7936	25473
(vi)	Kerala	12569	35941
(vii)	West coast of Tamil Nadu	844	7796
B.	**Bay of Bengal**		
(i)	East coast of Tamil Nadu	22411	33616
(ii)	Andhra Pradesh	16607	31044
(iii)	Orissa	17066	23629
(iv)	West Bengal	9935	22862
(v)	Pakistan	37255	65555
(vi)	West coast of Sri Lanka	9026	13579
(vii)	East coast of Sri Lanka	4711	11702
C.	**Mayanmar Sea and Malacca Straits**		
(i)	Mayanmar	108402	217500
(ii)	West coast Thailand	20961	58060
D.	**Islands**		
(i)	Andamans		15328
(ii)	Nicobar		728
(iii)	Laccadives		4436
(iv)	Maldives		7454

There are several types of devices to catch prawns, like nets, traps and hooks. Prawns are bottom dwellers, crawling over the bottom surface by means of their pereopods. But they can swim by their pleopods to move from place to place and in search of food. If disturbed they can spring from the surface and even jump over water. Light can attract them. Prawns are omnivorous, feeding on living animals and plants and also dead tissues. Their olfactory response leads them to decaying matters. Sexually matured prawns (penaeid) are found in the sea, whereas many of their juveniles migrate into shallow estuaries and backwaters to grow near adult size before going back to sea again. All these behavioural traits of prawns are taken into consideration in making nets, traps and other devices to capture them.

Cotton, hemp, nylon or in combination are used to make nets of various types. The simplest type is the cast net operated by a single man throwing it over the water surface. Bag-shaped nets (bag nets) are tied to the stakes and made stationary (stake nets). In these nets prawns get entangled. Bag nets can be pulled by moving crafts, when they are known as drag or trawl nets. In some of trawl nets the upper margin of the mouth is fixed to a beam (beam trawl) with lower margin hanging free. In others, the net is provided with wings and attached to otter boards, called otter trawls. In tidal areas, nets without bags are tied to stakes for a long distance to prevent fish and prawns from escaping along with the receding currents. Such nets are known as barrier nets.

Nets which are suspended down in water by attaching floats on the upper margin and sinkers in the lower margin are the seines. A stout rope, the head rope is attached to the upper margin, while a ground rope or foot rope passes along the lower margin of the net. In gill nets, which hang in the water, fishes are enmeshed by their operculum; while prawns, lobsters and crabs get entangled by their spiny surfaces and appendages. Seine nets may have bags, and when made to move by towing them on to boats they are called boat seines. Shore seines are those which are laid in a circle near the shore and dragged to the shore. When nets for gilling are allowed to drift along the current, they are called drift nets.

The Chinese nets are very large dip nets supported on long poles. When they are operated at night, lighted lamps are used to attract prawns. Several smaller version of these nets in use are the scoop nets and hand nets.

Traps are of different types like the plunge basket trap and the bait-attracted trap to capture prawns.

Fixed Bag Net

Bigger 'dol' net and smaller 'Bokshi jal' are extensively used all the year round along Maharashtra coast to catch fish and prawn. They are made of either cotton or hemp.

Dol is conical in shape, size variable from 12 to 200 m long, from mouth to cod end, mouth 5 to 90 m in circumference with mesh size 4 to 12 cm near mouth reducing gradually to 1.0 cm at cod end. The entire net is made up of five pieces from mouth to cod end. A removable bag is attached to the cod end keeping the mouth open by a small pole, which acts as a trap for predatory fishes.

Dol net is set against the current of either high or low tide, tied to two poles of about 40 m driven into the muddy sea bed, protruding about 3 m of the pole above the water. The net is tied to two loops fixed on the poles, the mouth remaining wide open. When the tide fluctuates, the mouth is raised or lowered. No floats above and sinkers below are used in this net. A series of nets are tied in a line and hauling is done by 4 or more boats. The current sweeps the catches into the net and are emptied when the tide turns. The dol net fishes a few fathoms above the bottom.

The bokshi jal is also conical, but smaller than dol, average length 10 m, mouth 2 to 3 m in diameter and cod end is about 12 cm in diameter. This is also operated in the same way in the shallower regions, whereas dol net is operated in deeper waters.

Prawn that are caught in the net are *Metapenaeus affinis, Parapenaeopsis hardwickii, P. stylifera, Acetes indicus, Atypopenaeus stenodactylus, Hippolysmata ensirostris* and *Palaemon tenuipes*.

The dol net operating in *Saurashtra* coast is also a wide mouthed conical bag with mesh size varying from 22.86 to 30.48 cm in mouth, gradually decreasing to 1.25 cm in cod end. Carvel built boats are used to operate this nets. Along with the force of tidal current, fishes and prawns are engulfed within the net and emptied with the current turns.

Dol net in Kathiawar region, made up of twisted hemp with coir rope margin, are large conical bag, 50 m long with mouth about 85 m in circumference. Marginal coir rope is 2.5 cm diameter. Cod end is made of thicker twine, in some made of double threads. The net is of one single piece, the mesh at the mouth is about 9.2 cm decreasing to 0.6 cm at cod end. The net is operated mostly during post-monsoon months till the commencement of summer to catch prawns species, like, *Penaeus carinatus, P. indicus, M. monoceros, M. brevicornis, Parapenaeopsis sculptilis*. The net is set against the tide in muddy ground near rocky shores and is fixed with two wooden barrels as floats and two heaps of heavy stones as anchors, but in a simpler way. A rope usually made of palmyra leaf fibre is passed through stones on each side, kept 45 feet apart. The same ropes pass through the edges of the net and at the upper corners, the free ends are tied to barrels. Cod end is tied. Nets are hauled at the turn of tide. Prawns are caught along with fishes.

Long tapering bag, 15 m long with 18 m circumference in the mouth are used all throughout the year in the backwaters except in very heavy monsoon. The mesh size of the net is 20.5 cm at mouth, 11.5 cm immediately behind and gradually reducing to 3.8 cm and 1.28 cm at cod end. Four coir loops are made, one pair at the top and another at the bottom parts of the mouth. Two stakes are driven into the substratum and net is tied to them by means of the loops on the border of the mouth. The net is set in position against the current just before or soon after ebb tide and the operation is by tidal action as in dol net. The lower border of the mouth of the net touches or is very close to the bottom. Hauling is done when the tide turns. The time of operation is generally from evening to early morning.

The catches of *Metapenaeus dobsoni, M. affinis, M. monoceros* and *P. indicus* forming bulk of the catches appear to follow a pattern of rainfall, more the catches heavier the rains. Highest catches are obtained on new and full moon days or a day or two later, but rarely earlier. Strong tidal currents on these days may force a large volume of water through the net, capturing a large number of prawns.

In Andhra region of east coast, a long tapering bag of 13.7 m long, made of cotton and coir yarn are used in Godavary and Krishna Deltas. The mouth of the net is rectangular with a head rope 5.08 m, side ropes 3.8 m and foot rope 5.9 m long. Mesh size is 10 cm at mouth, gradually decreasing 1.0 cm at cod end. Nets are fixed to stakes driven in mud by tying with head and foot ropes above and below. The nets are set against current. The foot rope is longer than head rope giving a firmer hold on the stakes preventing it from lifting off. From the head rope two ropes with weight are suspended down to enable it to withstand the force of tidal current. 10 to 30 nets are operated in a row. When the tide recedes nets are hauled. *Metapenaeus brevicornis, M. monoceros, Penaeus indicus* and *Penaeus monodon* are caught in this net.

Bag net with a wide mouth narrowing towards the cod end, known as Behundi jal, are operated in the estuaries of West Bengal in Hoogly and Matlah regions, are made up of hemp and coir rope at the mouth to catch prawns. like, *Metapenaeus brevicornis*, and *Palaemon styliferus* in large quantities from October to March. Other prawn species, caught by the net are *Palaemon tenuipes, Palaemon flumicola, Macrobrachium mirabilis, Macrobrachium malcomsonii, Macrobrachium carcinus, Macrobrachium lamarrei, Metapenaeus brevirostris, Metapenaeus monoceros, Penaeus indicus, Penaeus semisulcatus* and *Parapenaeopsis sculptilis*.

Behundi Jal have wings extending from the mouth. The bag portion is 20 m long, mouth 6 m wide with 9 m long wings. Mesh size near mouth is 4 cm decreasing to 0.5 cm at cod end.

Nets are set against the current by fastening the wings to a pair of heavy wooden anchors in deeper areas, or two wooden poles in shallower regions. The mouth of the net is kept open by two bamboo poles fastened to the upper and lower lips of the mouth of the bag. A wooden barrel or kerosene tin is tied to each end of the wing to serve as a buoy. The cod end is tied with a rope and a float is attached to it to know the location. A dinghy with 2 or 3 men is required to set the net. Hauling is done when the tide turns and after emptying the contents the net may again be set against the prevailing tide. In the lower regions of estuary, the cod end is hauled and emptied every one or two hours.

Near the shallower regions of the sea, nets are tied on bamboo stakes, the wings kept stretched by tying bamboo poles at right angles to the stakes.

The mesh size and the size of prawn caught in stake nets were observed as follows. The size of *M. dobsoni* in stake nets varied from 11 mm to 90 mm with individuals between 41 to 80 mm in length forming about 96 per cent. Those above 81 mm in length constituted only 0.6 per cent. *M. monoceros* ranged between 41 to 80 mm contributed 53 per cent and those above this length was 46 per cent. In the case of *M. affinis* the range in size was between 21 and 100 mm. Prawns between 41 and 80 mm formed 88 per cent, those below 40 mm forming 11 per cent and those above 80 mm 1 per cent. *P. indicus* was caught between 21 and 140 mm, of which 19.2 per cent were in the size range of 41-80 mm, 32.7 per cent were of below 40 mm and 48.1 per cent were above 80 mm. The stretched cod-end mesh size of the existing stake nets varied from 8 to 12 mm and of paddy field filter nets from 9 to 28 mm and gills nets from 30 to 55 mm.

Barrier Nets

In creeks and tidal inshore areas of Gulf of Kutch and South Gujrat, barrier nets without bag, made of cotton, are used to catch *Metapenaeus brevicornis*, *M. kutchensis* and *P. indicus* during August to December. The size of the net is variable, usually 12 to 30 m long, 1.5 to 2.5 m high with mesh of 1.0 or 2.0 cm. Several pieces of nets are fixed on stakes driven at intervals in mud. No floats and sinkers are used. The net is laid pleated and concealed in the intertidal regions. Its ends are tied to trees or poles fixed for the purpose. When the tide is almost full and water has risen up, fishermen wade through the water carrying stakes in their arms, fix them at intervals, lift the net and tie them to the stakes. When tide recedes prawns and fishes get caught in the net.

Barrier net extending over a mile, consisting of a number of rectangular pieces attached to bamboo poles of 2 m high is operated in wide stretches of mud flat, which are exposed at low tide in Andhra region, particularly in Kakinada Bay and back waters to catch prawn like *Metapenaeus brevicornis*, *Penaeus merguiensis* and *P. monodon* along with fishes.

About 30 men carry the net in 2 or 3 Navas to the mud flats during high tide just before the tide turns. The bamboo supports taken are driven into the mud and the net is fixed on them in about 1.6 m water. The water during ebb tide flows through the net and prawns and fishes get trapped in the net.

Seven metre long and 3 to 4 metres wide net pieces are joined together to form a barrier net in winter months in the fore shore areas of West Bengal during springs tides when large areas get exposed during low tide. The net made up of cotton or hemp or jute with 5 cm mesh in the upper region and 1.3 cm mesh in the lower region with stout ground rope of jute is fixed during low tide in the tidal area with the help of poles to catch prawn like *Palaemon styliferus, Macrobrachium lamarrei, M. rudis, M. Javanicum* and *M. scabriculum*. The length of the area of operation may be even 1600 m and the stout ground rope is tied to the poles at the base. The head rope may also be tied above in some places to facilitate the lifting of the net later. At the highest tide fishermen go in boats, raise the net and tie the head rope above the level of water. When the tide turns and water recedes, many prawns and fishes get entangled in the net and others trapped in pools to be scooped by hand nets.

The drgnet, when required is used as barrier net in Coromandal coast and Pulicat lake during December to February to catch *Penaeus carinatus, P. indicus, Metapenaeus dobsoni, M. monoceros* and *Penaeus monodon*.

The net is fixed in shallow regions depending on the tide and winds. In northern regions wind is taken into consideration and fishing is done only in few places. In the southern parts tide is the main factor and more fishing is done. The drag net is fixed on stakes thrust in the mud at intervals. One half of the net is turned in a helicoidal fashion and is made into an imperfect coral trap, while the other end remains straight acting as a leader to guide the prawns into the end trap. The operation is done at night.

Boat-Seines

Boat-seines, made up of cotton or hemp with strengthening ropes, are used in Kerala coasts especially during August to October, when prawns have a tendency to migrate upwards. The net consists of three parts, a conical bag of 8.5 m long with a wide mouth, a platform 11.5 m long and 53 m long wings on either side. Mesh size varies from 1.0 cm at the cod end to 2.0 cm at the platform.

The net is carried in two boats, with the net between, to the fishing ground. On reaching the fishing ground the two boats move away from each other in a semi-circular manner, paying out the net. The towing ropes on the end of the wings are tied to the boats. After moving for sometime, when the shoals get collected in the net, the boats come closer, enclose the shoal and the net is hauled. *Metapenaeus dobsoni, M. momoceros Parapenaeopsis stylifera, Penaeus indicus* and *P. monodon* are caught in the net.

In south-west coast (Vizhinjam), boat seines, made up of cotton and coir yarn are used from April to October to catch *Metapenaeus monoceros* and *Parapenaeopsis stylifera* along with fishes. The conical portion of the net has a depression in the middle, and is made up of 4 pieces, the height and mesh size of these sections respectively are 1.8 m / 0.64 cm, 1.2 m / 0.64 cm, 5.6 m / 0.64 cm and 0.15 m / 0.35 cm. The wing on either side of the net measures 27.7 m in length with 35 meshes breadth-wise, each 22.4 cm knot to knot inside.

Two catamarans or boats are used for operation. Net is loaded in one boat and two persons paddle along. After reaching the fishing ground, they paddle away from one another paying out the net, each holding one end of the warp. After working the net around a shoal, they come closer, meet and haul the net.

Bag like net, made of cotton without wings are operated in central region of Kerala coast to catch *Penaeus indicus, P. monodon, Metapenaeus dobsoni, M. monoceros, M. affinis* and *Parapenaeopsis stylifera*. Floats are attached to the head rope and sinkers to the foot rope. The net may be 50 to 60 m long and 30 to 40 m at its maximum width. Long ropes are tied on either side of the net.

A plank built boat with a dozen of men are required to operate the net. The net is taken in the boat and is payed out, one rope being held by men in the boat, while the other end is held by one man who jumps into the water along with the rope and keeps swimming with it. In the beginning the boat and man move wide apart for spreading the net. On properly spreading and the mouth wide open, the boat and the swimmer gradually move towards each other and finally the swimmer enters the boat with the rope. The net is then dragged by the men and finally closed and hauled in the boat. The net is used in the Chakara region in Kerala during south west monsoon.

The nets similar in shape and construction (boat seines with bags) are used in Orissa and Andhra Pradesh to catch *Penaeus indicus, P. semisulcatus Metapenaeus dobsoni, M. monoceros* in Orissa and *Metapenaeus affinis, M. dobsoni M. brevicornis* and *M. monoceros* in Andhra Pradesh.

The net is a conical bag, the size of the big one is 13 m long with a wing span of 27 m on each side. The small net has a bag length of

7 m and the wing span of 15 m on each side. The mesh size is 9.0 cm near the mouth decreasing gradually to 1.0 cm at cod end. Head rope has floats and foot rope sinkers. The cod- end is tied with rope and a weight is attached to it to keep the net at proper level during operation.

Two catamarans are required for the big net with 2 men and one catamaran for the small net. The net is laid out like other boat-seines and after pulling the net for some time the catamarans come closer and the net is hauled.

In the inshore regions of Tamil Nadu, *Penaeus indicus, P. semisulcatus P. carinatus, P. monodon, Metapenaeus dobsoni* and *M. monoceres* are caught in a conical bag net, made of cotton and hemp, that resembles a trawl net in general shape. The bag is made of cotton and the cod-end of hemp. Mesh size is 30 mm at the mouth decreasing to 12 mm at cod-end. Two wings are on either side of the bag having a length of 21.4 m. The total length of the net is 34 m including the wing. Large wooden floats are attached on the head rope. Sinkers are not used.

The net is dragged by two catamarans. The net moves close to the sea bed even though no weight are attached to the ground rope.

Shore Seine

Shore seines in Karwar and Konkan coasts, made up of 100 to 600 rectangular pieces joined end to end, is operated in inshore regions in pre-monsoon months, to catch *Metapenaeus affinis, M. dobsoni, Penaeus indicus* and *Parapenaeopsis stylifera* along with fishes. Each piece varies from 2 to 6 metres in length and 5 to 11 metres in height. Mesh size of side pieces is 3 to 5 cm and that of middle pieces from 1.2 to 2.0 cm. Floats on cork-line and sinkers on lead-line are fixed. At the ends long ropes are attached. Catches are collected in the 'bunt' of the net.

To operate a full-sized net, 60 to 80 men and 4 or 5 boats are required. The free end of the net is held on the shore pulled by 20 to 25 men, held in such a way as to open the net without collapsing. Net kept in V-layers in big boat and rowed perpendicular to the shore with 16 to 20 men. The boat with net steers in a semi- circle and brings the other end of the net to another group of men on the shore. The boat then rows back and anchors in the centre of the net. The net is dragged to the shore by the two groups of men and it comes out of water. When the ends are about 150 m apart, dragging is stopped. The portion of the net dragged out of water on either side is tethered to crutches fixed in the sand above high water level and the tow ropes are rolled up on stones. The foot rope of the portion of the net in the breaker area is weighted down by a few stones to prevent rolling over. As an additional measure to keep the net in position, head rope is tied to a few small boats. Foot rope beyond breaker level trails the ground. In this way prawns and fishes are impounded.

Smaller shore-seine of 20 to 30 pieces or 50 to 60 pieces, each 4m long and 5m high is also operated during monsoon periods from June to September and also from March to May with the help of one boat and 8 to 12 men.

Shore-seine made of cotton is used from November to March and some throughout the year in Andhra coasts. The central portion of the net is with 3 to 6 rectangular pieces, each 7 to 8 m high and 4 m wide. On either side, long tapering wing is present with 20 to 33 sections of netting, each section being 18 m high nearer the centre, the height decreases towards the free end to 1.1m. At the free end there is short bamboo pole of 0.55 m height to which is attached a hauling rope. Wooden floats are attached to the head rope and the sinkers on foot rope at intervals of 3 m. Five large wooden floats are attached to head rope, one in the centre and 2 on wings by which the position of the net can be known. Mesh size is 1.2 cm in the centre and 4.0 cm at the end of the wings.

Net is carried in one boat and is laid in the sea, one end being caught by men on shore. After rounding the other end also is given to men on shore. The two ends are pulled and the net is dragged to the shore and brought together. When finally drawn on shore the central portion folds and forms an effective bag. Prawns, like *Metapenaeus affinis, M. brevicornis, M. dobsoni, M. monoceros, Penaeus merguiensis Palaemon tenuipes* and *Acetes* spp are caught in the net.

A funnel-shaped shore-seine with bag and two wings, made of coir and cotton are operated in inshore areas of Kerala coast during October and May to catch *Metapenaeus monoceros, M. affinis* and *Penaeus indicus* along with fishes. The cod end of the net is 7.6 m long and the remaining portion of the bag is of 8.3 m in length. Mesh size is 0.76 cm at the cod end which increase gradually to 1.7 cm at the mouth. The cod-end has a breadth of 2.7 m. The wings made of coir have a length from 305 to 610 m and attached along the lateral margins of the remaining portion of the bag. The mesh size of wings varies from 15.2 to 22.9 cm near the bag portion and 60.9 to 91.4 cm at the extremities. The warp is of coir attached to the wings on each side and is about 215 m long or more.

The net is taken in a boat, one end held by men on the shore. After laying the net the boat rounds and comes to the shore and hands over the other end of the net to another group of men on shore. When the net is dragged close, 3 or 4 men jump into the enclosure formed to beat the water and scare the prawns and fish inside. After the two ends meet the net is hauled.

As a result of upwelling when prawns are driven closer to the shore, in Tamil Nadu, Andhra and Orissa coasts, shore-seines with bag made of cotton, hemp and coir are occasionally used to catch *Penaeus monodon* and *Metapenaeus monoceros*.

Similar to the shore seine of Kerala coast, except that the wings are in two parts, a proximal 1.5m long hemp portion and a distal 300 m long coir portion, the net is funnel shaped 8.5 m long, 18.3 m broad at mouth and 3 m broad at cod-end. The cod end is long truncate and detachable 7.3 m long. Mesh size is 4.5 cm at mouth and decreases to 1.2 cm at the cod end. Wings have 10 cm mesh at bases and 50 cm at extremities. To the free end of each wing hauling rope of 1 km length is attached. Wooden floats are attached to head rope at regular intervals, right upto the wing. Sinkers are not provided to the foot rope.

Masula boats carry the net and lay it out in a semicircle. Each end is carried by 12 men on shore who drag the net as in the usual shoreseine and haul it.

Drag nets, a plain strip of net to which sticks are attached at intervals are employed to catch prawns in shallow areas and backwaters with slight changes in west and east coasts of India. The length of the sticks will be less than that of the breadth of the net. When the net is dragged the force of waters makes it a baggy into which fishes and prawn enter.

In Tamilnadu, cotten is used to make the drag net. The net is big, usually 18.3 m long with 30 to 40 spreader sticks, 0.7 to 2.75 m in height. Smaller nets are 3.6 to 4.6 m long with spreader sticks 0.6 to 0.7 m high, Meshes vary from 0.6 to 1.3 cm.

Two men drag the net in water, wading in shallow regions. The net becomes baggy when pulled through the water and prawns form a good part of the catch along with small fishes. *Metapenaeus dobsoni, M. monceros, Penaeus carinatus* and *P. indicus* are caught, mostly in juvenile stages.

In the creeks of Godavari delta, of Andhra Pradesh drag nets of 7.9 to 9.1 m long with width increasing from 5.9 m at the one end to 10.4 m at the other are used to catch *Metapenaeus monoceros, Penaeus indicus, P. monodon* and *Macrobrachium* spp. The mesh at broad end is 3.0 cm gradually decreasing to 1.2 cm at the narrow end.

One boat and two men operate the net. The narrow end of the net is secured to the gunwale of the boat by tying ends of the net to bamboo poles fixed on the sides of the boat. Two bamboo sticks 0.55 m high are attached to the corners of the broad end and 2 men drag the net along with the boat towards the shore by wading in waist-deep water. On reaching the shore the broad end is lifted and catches are shifted to fall into the boat.

During monsoon in Saurashtra coast, drag nets of 5.5 m long and 1.75 m high with 1.0 cm mesh fitted in bamboo poles at the end of the net are used to catch juveniles of *Matepenaeus kutchensis, M. monoceros, Penaeus carinatus* and *P. indicus*. Fishermen keep the net stretched between the poles and drag the net in waist deep water.

In shallow areas of backwaters and estuaries along Kerala coast, drag net made of cotton are used to catch *Metapenaeus dobsoni, M. monoceros, Penaeus indicus* and *P. monodon*.

Six pieces of net, each 1.5 m long and 4.5 m broad, mesh size varying from 1.0 cm at centre to 2.0 cm at the extremities. 2 to 3 layers of coir mesh is attached to the upper and lower margins of the net on which head and foot-ropes are attached. About 13 sticks, 1 m high are tied to the head and foot ropes at 0.75 m apart. The net bulges like a bag when dragged in water.

Two men hold the sticks vertically at the ends and one holds the stick at the centre and dragging and moving along finally coming close together and hauls the net.

Some gills nets are made use of to drift over bottom surface to catch prawns. From November to April, bottom drift net, made of cotton, hemp or nylon are used in North Kerala and Kanara to catch big sized *Metapenaeus affinis, Penaeus indicus* and *P. merguiensis*.

Net made up of 16 to 25 pieces of net, each 3 to 5 m long, 2 to 3 m high with 5 to 6 cm mesh size is provided with floats in head rope and sinkers at foot rope at intervals to keep the net in position. Two to three men in a canoe allow the net to drift with the current on the bottom for some time, and lift the net for catches. Prawn get entangled in the net along with fishes which are gilled.

In the south west coast, surface drift nets, made up of 40 to 50 pieces, each 5.8x1.8 m of 3 cm mesh size joined together and are operated by two men in a Catamaran, all most all the year round to catch. *Penaeus indicus* along with fish, *Chirocentrus* spp.

In the Andhra coast, similar drift nets are operated by 4 to 6 men in a boat for most parts of the year. Large-sized prawns, *Metapenaeus brevicornis*, *M. affinis*, *Penaeus indicus*, *P. merguiensis* and *P. monodon* are caught in small numbers.

Falling Net

Falling nets (cast nets) are used in shallow areas of the coasts and back waters. Nets made of cotton or nylon are more effective on muddy or sandy bottom without any rocky projections to catch young ones of *Macrobrachium* spp, *Palaemon* spp, *Metapenaeus* spp and *Penaeus spp*.

One man operates the net from the shore if the fishes and prawns are nearby, or from a canoe in the creeks and back waters. The net is taken, a part of it in each hand and thrown in the air with a strong swing to spread it and it falls on the water and sinks down covering the prawns and fishes, if any within its circumference. The central string is held in hand and the net is slowly pulled till the edges close.

Conical in shape, the net form a circle when spread out. The foot rope along the circumference of the net has small lead weights attached to all round about 15 to 20 cm intervals. Each weight is cylindrical, 3 cm long, 1 cm diameter at the centre. A string passes from the centre which is held in the hand of person for operating the net. The net have a radius of about 2.5 to 3 m with 1.0 to 2.0 cm mesh size.

A heavy large sized cast net of 15 to 23 m diameter, the edges with puckerings and sinkers is operated from a long narrow boat with 5 men in West Bengal. Net is kept in full length on one side of the boat and dropped as it drifts with the current enclosing a circular space, and when the central string tied to the boat is taut the net is hauled up. Prawns and big fishes are caught in the net.

In Kerala and Kanara coasts scoop nets consist of a bamboo pole of 2.5 to 3.8 cm in diameter is bent like a U, the free ends tied with a thick rope or hemp string. The free ends are 0.9 to 1.2 m apart serving as handle of the net. A net of 0.6 to 1.2 cm mesh is fixed on this frame, which remains bag like, the depth being 1 m. One man holds the frame with both the hands and wades through water, diping, pushing and lifting at short intervals to catch small amount of *Mepatenaeus* and *Penaeus* spp juveniles and small fishes.

Juveniles of *Metapenaeus* are caught from estuaries and creeks of. Andhra Pradesh in a bigger scoop net by pushing. The net is 1.8 m long and bag-like, the cod end being truncate. Mouth is fixed in a triangular frame, whose base is of flat wooden plank 1.2 m long, and sides are of bamboo poles 1.4 m long which cross at the apex and project beyond for 0.5 m serving as handle of the push net. Mesh size varies from 1 cm at mouth and 0.7 cm at the cod-end. One man pushes the net by holding the bamboo handle through shallow areas, wading through water. It is lifted at intervals to empty the catch.

Scoop nets made of cotton and jute are operated in Godavary estuary and Kakinada Bay during day and night when the tidal current is strong to catch *Metapenaeus brevicornis, M. dobsoni* and *Penaeus indicus.* Better catches occur in night time, when prawn form 50 to 70 per cent of the total catch.

Length of the net is 10.2 m, one end narrow (13.3 m broad) and the other broader (76.9 m broad). The mesh size at the wide end is 6 cm narrowing gradually 2 cm at narrow end.

A boat with 4 men operate the net. Two men in the boat hold the narrow end of the net above the water. Near the wide end two triangular platforms are erected on each of which two men stand with a pole in the hand of about 2.5 m long. The lower end of the pole is tied to the corners of the broad end of the net. When tides are flowing inwards the two men lower the net with poles under water to the bottom of the creek, the net forming a shallow bag through which water filters. Every two or three minutes the net is lifted high above the head of the men holding the poles and catches are passed to the narrow end of the net from where they are removed by small hand nets into the boat.

Lift Net

Penaeus and *Metapenaeus* spp are caught in lift net (Chinese net) in Kerala. The size of the net is variable according to the size of frame work set up to operate the net. Smaller nets are operated in the interior of back waters, while at the mouth of the harbours bigger nets are used. The net is a square piece of about 10 m on each side. There is a frame work for propping up the net. Two stout wooden piles are driven in water about 1.8 to 2 m apart, the upper portions rising above high water level. On each of these a hole made and one stout piece of wood with ends narrowed is fixed, not very tight in the holes. This acts as the axle and can be rotated. From near the two edges of this axle two long narrow poles about 9 m long pass inwards and their ends are fixed together forming a triangle. Similarly two poles stretch upwards in slanting manner, their ends also joined as an apex of a triangle. The inner and outer apices of these are joined by a few bamboo poles joined together. From the apex of the upward poles four curved poles are fixed, which are directed downwards like the feet of a spider. At the ends of these poles the four corners of the net are tied. The net hangs like a big wide mouthed bag. When not in use the cod end is pulled upwards and tied by a string. At the apex of the inner poles a number of thick ropes of different lengths are tied and on their free ends heavy stones are tied.

Two to four persons are required to operate the big net. The inner poles are given a push up, and when they rise, the axle rotates and the upper poles with the net lower down, and the net and part of the poles supporting it dips into the water. By pulling the ropes with stone weights the net is lifted again. After lifting by pulling the cod-end to which string is attached prawns and fishes inside it are scooped up by a hand net. At the harbour mouth big fishes are also caught. To attract more prawns into the net at night a light is kept hanging from the apex of the poles.

Traps

Several types of traps are operated in different regions of India to catch prawns, lobsters, crabs and fishes. Some traps are worked by hand to cover the animals from above, others are used to trap them while passing through the currents. In some kinds of traps, baits are placed to lure the animals to enter them. In some regions major portions of prawns and lobsters are caught by trapping.

Figure 35: Penaeid Prawn Landing from Trawler's Catch

Figure 36: Harvesting of *P. japonics* by Traps

The cover basket is used all over India for catching prawn and fishes of different kinds. This is a conical basket open at both ends, made of bamboo strips, or cane or narrow reeds laced together by coir rope all around at intervals. The opening at the top is 15 cm in diameter and the bottom is 0.45 m in diameter. The height of the trap is 0.45 to 0.6 m. The ends of the strips at the wide opening is sharpened for fixing them in the mud bottom. The sides of the narrow opening are bound by a few layers of coir rope to form a thick ring. One man carries the basket in hand, slowly wades through the water and plunges it into the water where prawns or fishes are expected to be present, and firmly presses the basket down in the mud entrapping them. By putting one hand through the top hole the animals inside are searched for and taken out.

Bamboo strips, 1.5 to 2 m high are laced together by coir ropes at intervals to make the screen by which a kind of cage is constructed in the tidal areas to catch prawns and fishes. During high tide they are fixed vertically by thrusting one edge into the mud as cages of different shapes, round, oval, rectangular. The place where their edges meet are adjusted in such a way that it forms only as a one way entrance. When water ebbs out at low tide the praws and fishes are stranded inside and collected.

Raft trapping of prawns are practised in the shallow areas of the back waters and their connected canals. Two medium–sized wooden boats are connected together by tying bamboo poles across both at the anterior and posterior extremities. After tying the distance between the boats will be about 1 m. Inside the boats weights like stones and sand bags will be kept on the inner sides so that these edges will be very near the water surface. A long iron chain is connected to the boats, its middle region lying in water touching the ground. This acts like a scare line. The boats are moved forward by punting with poles or rowing with paddles and when the chain is dragged at the bottom, it scares the prawns which jump out of water and fall into the boats. Inside the boat, twigs and leaves and a large number of empty fronds of coconut inflorescence are kept which prevent the escape of prawns by jumping back into the water. The operation is done in the night. In northern parts of Kerala on the outer and anterior sides of the boats, nets are fixed on supports inclined outwards to trap the prawns that jump in that direction. Young ones of *Metapenaeus* and *Penaeus* spp are caught in moderate quantities.

A conical cage trap is used all over Kerala to catch prawns. One end of this cage is narrow, the other end wide with valve-like arrangement for entrance of the prawn with no escape. The cage is made of split bamboo pieces 0.76 to 0.91 m long with the mouth 0.3 m diameter. The mouth opening is reinforced with bamboo pieces in the shape of an apron. The bamboo splinters at mouth project freely into the interior making a one-way entry valve. The splinters joining at the narrow end of the edge are tied together, which can be opened to take out the catches. The cage is placed in narrow channels and entrances into fields against the flow of water. The two lateral sides of the cage are bunded off with earth.

Baited float is used in the shallower areas of Kerala back waters. The visceral mass of snail is tied to a piece of strong string and at the opposite end a small float is tied. Several of these baited floats are kept in water in a small area where prawns abound. After some time the prawns begin to grab at the bait which is visible by up and down motion of the floats. The fishermen waiting nearby in small canoes approach and entrap them by cover basket. In the prawn season of June to October several hundreds are collected in few hours.

There are 67 areas leased out in Chilka Lake in Orissa where prawn trapping operations are carried out besides large number of unleased areas. *Penaeus indicus* is the most abundant species caught, followed by *P. semisulcatus, Metapenaeus affinis, M. dobsoni,* and *M. monoceros.* The bigger trap *Dhaudi,* is operated by Kandra class of fishermen and the *Baja,* the smaller type of trap operated by the Tiara class fishermen from March to end of June.

Daudi, a box type trap is 1.2 m high, 1.15 m long and 40 cm wide at bottom. It is made of bamboo splinters, those of the four sides tied together to a bar giving a roughly conical shape of a roof. There are two or three openings on one side having thin bamboo strip intersections through which prawns can enter but cannot escape. These traps are set in the lake partly submerged, the openings always under water. They are fixed to the bottom with long bamboo poles. A group of traps 5 to 10 in numbers are kept in a circular manner leaving a narrow opening as an entrance to the circular enclosure. From the centre of this gap a bamboo screen extends outwards of 180 to 275 m long, reaching upto the shore. This is known as leader line. Prawns which strike the leader line follow the line and enter the enclosure and later in the trap through the openings. In some areas hundreds of traps are kept without a leader line.

Baja, the smaller traps are set in a special way. Bamboo screens are erected in the lake to enclose an almost equilateral triangular area. The screens leave a gap at each corner, and in each of these gaps a trap is set, their entrances facing the enclosure. The opening of the trap is provided with a longitudinal valve-like arrangement through which prawns can enter but cannot come out. In the enclosure from a gap in the base, a long bamboo screen is set as leader. Prawns follow the leader line and following the screens of the enclosure finally enter the traps. The top of the trap can be opened to take out the catches.

Bamboo basket traps of hut-box type are used in the deltaic areas of Godavary and Krishna rivers of Andhra Pradesh and in the irrigation channels connected to the backwaters, during July to October to catch *Metapenaeus monoceros, Penaeus mondon* and *Macrobrachium* spp. Traps are rectangular in shape, made of bamboo

**Figure 37: Fishing Trawlers in Quest of
Marine Prawns and Shrimps**

**Figure 38: A Trawler Catch of Penaeid Prawn,
off Orissa Coast**

Figure 39: A Trawler Catch of *Penaeus monodon*, off Orissa Coast

splinters 1.5 x 1.0 x 0.7 m in size. The bamboo splints from the sides coverage to a point which can be opened and closed for emptying the catches. On the sides of the trap one or more valve-like openings are made for the entrance of prawns.

Net trap is operated in Pulicat lake, Tamilnadu, where prawns are led into a net. The net is bowl-like, made of cotton, 18.3 m in circumference and 5.3 m deep, with a narrow cod-end of 2.5 cm mesh. Two long coir ropes with small stones tied to it at intervals are attached to one edge of the mouth. The net is operated in deeper areas. One edge of the mouth of the net is fixed to the ground by stakes and the opposite edge is held above the water keeping the mouth open. The stone-laden tickler ropes are pulled out, and this disturbs prawns and flat fishes which move towards the net and finally enter inside. When the rope has come near mouth of the net the foot rope is lifted up and net closed.

In the rainy season, mainly for prawns, conical cage trap, made of fine reeds, are operated in the Ganga river system. It has a wider mouth at one end having 24 cm diameter, and narrowing to the other end to length of 1 to 2 m. The narrow end is tied with rope, and can be opened and closed when necessary. At the open mouth end reeds are fixed all around, their free ends converging inwards to form a one-way entrance. Prawns and fishes can pass inside the cage cannot escape. During operation, 4 to 5 cone traps are kept immersed in water one behind the other with some space in between each one. On either side of the open ends, barriers of dried weeds are fixed by means of bamboo sticks. The open end face the currents. The traps are fixed in the evening and opened in the morning and catches are taken out by opening the conical end.

Trawl is a very effective gear for capturing bottom-dwelling marine prawns. They are conical bag-like nets attached to moving boats, and while moving, the mouths of the net are kept open. In the beam trawl mouth is kept open by fixing the upper lip of the net on a beam which is supported on iron trawl heads. In the otter trawl there are two solid wood or iron otter boards in which warps are fixed and when in motion they tend to diverge as in a kite and keep the mouth of the net open. The otter trawl can be led only when the towing boat is in motion and hauling is also done when the boat is moving. Both the nets are conical bags which narrow at the cod-end, a cylindrical portion in which the catch is held.

The beam trawl is not very much used for commercial fishing now-a-days. It is mostly employed for surveying the bottom fauna.

Net in otter trawl is a conical bag with a tapering cod-end as in beam trawl net, but with differences. The square tapers to the lower side touching the anterior edge of the baitings, which tapers towards the cod- end. At the anterior most extremity there are two upper wings, one on either side. On the lower side there are two lower wings arising from the anterior margin of the belly, reaching upto the level of the ends of the upper wings and hence these are very long. On the inner margin of the top wings the head rope passes carrying a number of floats for buoyancy. On the inner margin of lower wings, the foot rope passes on which sinkers are attached. The head and foot ropes in either side are attached to the rear end of the otter board. The otter boards are of different sizes according to the size of the net. About a third of its length from one edge there are two brackets on the board, one smaller than the other. At the distal ends of the brackets, where they join a warp is attached. The board takes an angle to the direction of the tow and acts as a kite. The wooden board is weighted by nailing iron pieces to it.

There are two types of otter trawls. In one type the square portion extends forward forming an overhang over the foot rope, (over-hang type). In the second there is no overhang. In some cases there is a piece, called throat connecting the belly with the cod end. A small piece of webbing (batings) connects the belly with the cod end or the belly with the throat. Otter trawl is used for bottom fishing for prawns. The size of the otter board varies from 0.75x0.4 to 1.0 to 0.9 m with 11 to 90 kg in weight according to the size of the net and towing power required. The distance between the ends of wings varies from 7 to 27 m. Floats used on the head rope are spherical aluminium ones. The foot rope is longer than head rope carrying spindle shaped sinkers. Mesh size at cod end is 5 to 10 cm. Trawling speed ranges from 1.5 to 4.5 knots per hour according to the size of the gear. Tickler chain is also attached ahead of the mouth of the net to dislodge the prawns from the ground. With the small boats the trawl net is operated in the inshore waters upto a distance of 15 to 30 km. Exploratory trawling at 300 to 400 m depth yielded deep sea prawns *Penaeopsis rectacuta* and *Metapenaeopsis philippi*.

Bottom-set entangling net (Trammel net) is used in shallow coastal waters to catch prawns, mainly *Penaeus indicus*. The net is a three-layered one, with a fine net of smaller meshes (20 mm bar) hung loosely between vertical walls of coarser net of much larger meshes (135 mm bar), so that prawn passing through the outer wall carry some part of the finer net through the wall of the other side and are entangled in the pocket thus formed. The inner wall has 4500 horizontal and 72 vertical meshes and the outer walls have 583 horizontal and 8 vertical meshes. The webbing in both are rhomboidal. Polyethylene ropes (4 mm dia) constitute the float line, the sinker line, the buoy rope, the pull rope and the mounting lines which form an integral part of the net. The head rope is formed of two ropes, the float line and the mounting line. Similarly the foot rope is also comprised of two ropes, the sinker line and the mounting line. The floats are synthetic and round and the sinkers are lead and barrel shaped. The inner netting is hung from the mounting line of the head rope, where as the outer walls are tied at their upper and lower extremities to the inner walls two to four meshes away from the mounting lines. The floats are passed through the float line. The mounting and float lines are rigged by rigging twines at intervals of 20 cm and 40 cm alternately, with one float in each 20 cm interval. Similarly sinkers have been fixed on the sinker line by rigging the mounting and the sinker lines at regularly repeated intervals of 2 cm and 18 cm with one sinker in each 2 cm interval. Two granite stones each weighing 0.5 kg are tied to each end of the foot rope to anchor the net in position. A marker buoy is attached to the buoy rope which is the continuation of head rope. This is used to locate the position of the gear in operation. The proximal ends of the head and the foot ropes are continued 3 m from each side, united at ends and prolonged further as the pull rope.

The net is operated from catamaran of any length by one or two persons at depths upto 35 m.

Difference in size of prawns caught in Chinese dip net, stake net and cast net have been observed in Cochin backwaters. The mesh size of stake net from knot to knot when stretched was 85 mm at mouth decreasing to 10 mm at the cod end. In Chinese net the mesh size was 20 to 25 mm at mouth decreasing to 10 to 15 mm at cod end. In the cast net, however, mesh size was 28 mm near the edge lowered to 20 to 25 mm towards interior. The largest sizes of prawns were in the Chinese net catches followed by cast nets. Stake net catches were the smallest size. The difference in catches may be due to mesh size, mode of operation and behaviours of prawns.

The Chinese net is dipped and lifted from time to time, and is operated at night by using a powerful light to attract prawns. The prawns that move in the night for food or along with the current are attracted by the light and the bigger ones are most likely to be caught. In the cast net a column of water is covered by the net and prawns available in that area only is caught, moving or buried. In stake net catches the prawns are brought by current and some bigger ones may jump over the net or swim against the current away from the net. But in monsoon when currents are fast, bigger ones are also trapped. It has been observed that the size composition of *Penaeus indicus* changed with increase in mesh size of these nets.

In an experiment of luring prawns and fishes in Chinese dip net in Kerala back waters, it has been found that with increase in light intensity, 200 to 600 per cent increase in catches were obtained upto a maximum of 200 watt, after which the catches are reduced. Green, blue and red coloured lights were more effective than white light in attracting prawns and fishes, green being the most effective. Large specimens of *Penaeus indicus* (10-17 cm) were generally found in coloured lights, whereas smaller specimens (3-6 cm) were predominant in white light. The species attracted to lights were prawns *Penaeus* and *Palaemon*, Crabs, *Scylla* and *Neptunus* and fishes, *Mugil, Hemiramphus, Caranx, Arius, Equula, Stolephorus, Chaetoessus, Brachiurus* and cuttle fish.

It has been found that 21 per cent of prawns and 11 per cent of fish escaped beam trawl during operation. When tickler chain was attached to the beam trawl 30-48 cm ahead of ground rope, 47 per cent increase in prawn catches was obtained.

In otter trawl operations, an increase of 43 per cent catches in fully mechanized vessels with winches have been obtained over partly mechanized boats where hauling with hands have been done. On reducing the number of floats in head rope, prawn catches were more due to reduction of vertical height. A heavy foot rope and less buoyancy on the head rope is required to make shrimp trawlers more effective. Smaller trawl nets have relatively more catches than bigger ones as the smaller ones dragged more near the seafloor increasing its effectiveness for shrimps. For the effectiveness of prawn catches the nearness of net to the sea bed is more important than wings which may be much above the ground. By attaching a tickler chain and by reducing number of floats on the head rope there was an increase of 70 per cent of shrimp catches. Maximum effect was when chain was attached to the ground rope with less floats. Prawn catches were higher in cotton netting. Cotton being heavier than nylon may drag closer to the ground disturbing the prawns resulting in more catches.

Behaviour of Marine Prawns

Metapenaeus affinis and *Parapenaeopsis stylifera* exhibited electronarcosis and fixation in different times when exposed to current density of 1.2 ma/mm^2 at temperature and water resistance of 30°C and 750 ohms/cm^2 respectively, the periods varying inversely with the length of animals. *M. affinis* required 2.4 seconds to bring out narcosis and fixation in animals between 132 to 90 mm length in interupted A.C. of 37 pulses per minute when the body voltage between their head and tail were 9.6 to 6.7 volts. It was further observed that the effective period for narcosis and fixation of *M. affinis* varied inversely with the voltage tension between head and tail of the animals. The period of recovery of animals increased when treated with higher pulses for longer effective period. Repeated electrical stimulations had no marked effect on the period of narcosis and recovery except in some cases where the same period was required for fixation in second to fifth stimulations.

Reaction of Prawns to Low Volt Direct Current

Penaeus indicus, Metapenaeus dobsoni and *Metapenaeus affinis* of 18.5 to 22.0 cm; 10.1 cm and 13.5 to 15.0 cm body length were found to jump in any position between the two electrodes at voltages below 12 (4.0-12.0). *P. indicus* in particular was noticed to jump as high as 60 cm. Larger size groups of *P. indicus* and *M. affinis* require low voltage (4.0-8.0 volts) for causing appreciable effect on them compared to smaller ones. Further the voltage required for causing this reaction (jumping) in the shrimps is lower when they face the cathode than that required when they face the anode. *P. indicus* requires only 4 volt to exhibit jumping reaction when it faced cathode, while on facing the anode, the species showed the jumping reaction in 11.5 volts. Thus the exact position of the shrimp in relation to the electrodes appears to play a vital role. When a shrimp jumps and falls in line with the path of the current, it is stunned. But if it happens to fall in any other position (45° angle or perpendicular to the path of current flow) no apparent damage is caused. In a slowly rising current intensity of the field the prawns, therefore, manoevour to orient themselves in any position other than in line with the path of current flow, the moment they perceive electric field to avoid being stunned.

Chapter 9

Spoilage and Preservation of Marine Prawns

Spoilage

The degree of spoilage of fish is influenced by a number of spoilage bacteria. Their proteinase becomes significant when the number approaches 10,000,000 bacteria per gram of fish. The microflora of crustacean shellfish seems to be similar to that of fish. Bacteriological studies of shrimp showed that whole shrimp examined immediately after emptying of the trawl net varied in bacterial count from 1600 to 1,200,000 per gram. At the end of two days storage, the total bacterial count on whole shrimp increase sevenfold; on headless shrimp fivefold. However, fish caught in unpolluted water carry no bacteria of public health significance on their surface and other external surfaces and the flesh and internal organs are sterile.

Prawns are usually transported as whole in wooden boxes by truck from the landing places to the processing plants. The duration of transportation varies from 10 to 48 hours. 1cing of prawns is usually practiced. A proportion of one part of ice to one part of prawn is used for short distance (10 to 20 hours) and one and half parts of ice to one part of prawn for long distance (more than 24 hours). At the plant, shrimp are unloaded and dumped onto the floor of the plant and exposed there in heaps. Sorting and beheading of shrimp are done on the floor in the same area. Beheading merely consists of breaking by hand the shrimp at the junction between the head and body. The head with the walking legs is discarded or converted to fish meal and the remaining section is retained. Beheaded shrimp are deveined, trimmed and sorted by the plant employees to remove any discoloured, decomposed or mutilated shrimp. During this operation, beheaded shrimp are washed 2 to 3 times with clean potable water in a washing tank. The next step after the culling is the size-grading of the shrimp. Mostly they are done by the hand on aluminium-covered table. After sorting, the shrimp are again washed a few more times with the clean potable water in another washing tank. Sorted shrimp are then arranged by size in the metal container and chilled water is added to allow the free water to form a layer of ice covering the shrimp. After the shrimp are solidly frozen in a contact plate freezer, the block of frozen shrimp is taken out from the container and wrapped in a moisture -proof material and then put in a wax-paper carton. The packaged shrimp are packed in master cartons and returned to the freezer for storage until removed for export.

The quality of the raw material influence the quality of the finished product. Beheaded shrimp either before or after transportation, were found to be of better quality than the whole shrimp. Beheading of shrimp at the earlier stage improve the quality significantly. Washing of shrimp in clean water reduce the number of bacteria considerably. Use of chlorinated water in washing can reduce the number of bacteria further.

The current commercial practices of handling shrimp still need to be improved. Although the quality of the product is generally acceptable, measures should be taken to improve it. It must be realized that for freezing the raw material that arrives at the plant should be as fresh as possible. It is not possible to produce a first rate frozen product from a lower quality of raw material. Therefore, maintenance of the quality of the product should start from the very beginning, on board the boat, if possible.

(i) Handling of Prawn on Board

As soon as the catch has been hauled abroad, the prawn should be iced and stowed in the holds in boxes. The shrimp should not be left on the deck for hours, exposed to the sun and also should not be stowed in bulk in the holds as the weight of the upper layer will crush the prawn of the lower layers.

(*ii*) Handling at the Landing Place

Soon after landing, the catch should be transported to the processing plant. It should not be left lying exposed to the sun. The investigations indicate that beheading the shrimp immediately after landing reduces the bacterial load. If practicable, the various step, such as, beheading, deveining and cleaning could be done at the landing place immediately after landing the catch and the prawn re-iced with fresh ice and transported to the freezing plants.

(*iii*) Handling at the Freezing Plant

After beheading and deveining the prawn should be washed in clean water, preferably the chlorinated water containing about 5 ppm of available chlorine. Higher chlorine content may confer an odour to the product. The plant premises, working tables, utensils and other equipments should be washed at the end of each working day with chlorinated water containing 25-50 ppm of available chlorine. The plant premises should be kept absolutely clean and refuse material, such as, the heads, viscera etc. should not be left exposed but should be put into covered containers. Measures should be taken to prevent contamination by flies.

Preservation

The share of prawn in marine product exports from India is approximately 60 per cent in quantity and around 80 per cent in terms of value. Sea food export industry of the country with a humble beginning of rupees 50 laks has grown to an organized industry, earning valuable foreign exchange of Rs. 8363.53 crores in 2006 exporting 612641 tonnes.

Prawns mostly in frozen, dried and canned forms are exported from India to different countries of the world. Among them Japan, USA, Australia and UK are the major markets for frozen prawns. The importing countries for the dried prawns from India are Belgium, Mauritius, SriLanka, USA, Germany and UAE and for the canned prawns are France, Netherland and USSR. However, for the frozen prawns, Japan accounts for the biggest share of 55.77 per cent in volume and 70 per cent in value during 1988-89, Japan used to buy both big sized shellon prawn and small sized peeled shrimp from India and major share being the small sized shrimp. Besides in the same year, for the first time India exported Individual Quick Frozen (IQF) prawn of 168.4 tonnes amounting to Rs. 163.5 lakhs to Japan.

All these facts, reveal that prawn is the most important commodity in the present day fishery industry to earn foreign exchange for the country. It has helped a lot for all the modern mechanized fishing and most sophisticated methods of preservation on it to come up in the largest way. Among different prawn preservation methods, most important are freezing, drying and canning. Besides, other diversified prawn products are dry prawn pulp, breaded prawns, prawn flakes, shrimp extract and prawn pickle. Prawn by-products which include prawn manure, chitin and chitosan are also very important from industrial point of view.

Freezing of Prawns

Freezing of prawns have made a considerable headway during last few decades and considerable improvement in the quality of the products have been achieved by employing newer and more efficient techniques of handling and processing. In commercial practice, several types of packs of frozen prawns have turned out for export. Basically these are based upon the method of preparation of the raw material for freezing. The different types of packs of frozen prawns are, (a) Headless shellon (HL), (b) Peeled and Deveined (P and D), (c) Fan-tail / Butterfly, (d) Peeled undeveined (PUD), (e) Whole/head on (e) Cooked peeled (CP), (f) Cooked, peeled and deveined (CPD) and (g) Peeled and Deveined cooked (PDC)

Flow Chart of Freezing the Prawns

Raw material → Washing → Dressing → Washing → Draining → Size grading → Weighing → Packing → Freezing at –40°C → Glazing → Packing in duplex cartons → Packing in master cartons and strapping → Frozen storage at –20°C

Considerations in Prawn Freezing

Frozen prawn product has become the vital item of the sea food freezing industry in India, because of its export value. The final quality of frozen prawn depends upon the freshness of raw material, handling of fresh prawn and the processing techniques to be followed. Some of the common defects occurring in frozen prawn

Figure 40: Grading of Prawn Prior to Freezing

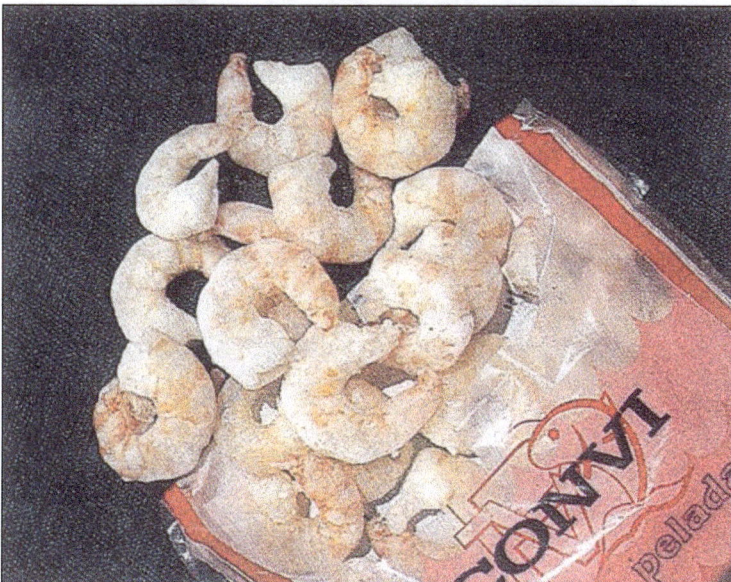

Figure 41: Processed Prawn, Ready for Sale

Figure 42: Prawn Freezing Plant

Figure 43: Frozen Block of Tiger Prawn

Figure 44: Testing of Quality Products

products are dehydration (Freezer burn) and discolouration of shell and meat. Under weight conditions are generally encountered in frozen peeled and deveined prawns as they hold large quantities of free water in fresh condition and the water holding capacity is reduced during thawing after freezing. Weight losses in frozen prawns generally varied from 7 to 12 percent in P and D, 5 to 7 percent in HL and 7 per cent in cooked and peeled prawns during storage for a few days.

Black spot formation or melanosis is chiefly because of the contact of prawns with atmospheric oxygen. This occurs by a reaction involving free amino acids, phenolase enzyme and oxygen, which converts phenols to melanin pigments and these pigments generally occur on the internal shell surfaces or in advanced stage on the underlying shrimp meat.

Another problem of loss of quality during the early period of storage is mainly caused by autolysis and with longer storage, spoilage occurs mainly through bacterial action. Prawns live only a few minutes after removal from their natural habitat. Microbial spoilage starts immediately through bacteria on the surface and in the digestive system through microorganisms which happen to contaminate the shrimp on the deck during handling and from ice, used during storage. Removal of heads reduces the bacterial count somewhat, because the head carries approximately 75 per cent of the total bacterial load. Washing of the raw material must be done frequently to reduce the microbial contamination to the minimum.

Driploss during thawing is another important problem in the frozen prawn industry. Generally, thaw driploss occurs to the extent of 5 percent in headless shell on, 10 per cent in peeled and deveined and 3 per cent in cooked and peeled styles of prawns. This can be alleviated by a dip treatment of the prepared raw material in an aqueous solution of 12 per cent sodium tripolyphosphate and 8.6 percent sodium dihydrogen phosphate or 2 per cent citric acid. Thus it maintains correct drained weight and improves the yield during prolonged frozen storage and also protects the frozen products from denaturation of proteins.

Quality Control of Frozen Prawns

Frozen prawns must be prepared from clean, wholesome and fresh prawns and there should not be any visible signs of spoilage. Colour, meat texture and odour of raw material must be having typical of freshly caught prawns. Standard size grading of prawns must be followed. The frozen block, if having any white patch dues to dehydration (freezer burn) or black spot must be noted carefully and removed. Thaw driploss of the prawns must be observed and taken care off. Any discolouration of shell and meat is to be noted quantitatively. Number of deteriorated pieces are to be counted and removed. Overall odour of the thawed material and the flavour of the meat need to be assessed. A pleasant flavour indicates the product is of prime quality. Presence of any extraneous materials like, legs, bits of veins, loose shells, foreign materials etc should be noted and get rid of. Totals bacterial plate count, counts of *E. coli* and

Enterococci should be determined before thawing. High bacterial count indicates the unhygenic handling of raw material.

Pretreatment

Pretreatment is necessary before the freezing and storing process depending on the kind of prawn and their size.

After removal of head and without removing the shell, the whole body is frozen. This is practised with small and medium sized prawn in blocks. In case of large sized prawn (tiger prawn), individuals are freezed quickly and seperately for IQF products. In case of broken shelled prawn, shells and veins are removed before freezing the whole body enblock to produce P and D (peeled and deveined) product.

If wrongly applied, freezing spoils the quality of prawn meat. When the prawn is frozen, some physical and chemical changes occur in the prawn meat. So prevention of them is important. The changes are, swelling of the muscle fibre in the meat by formation of small ice crystals which damages the quality of the meat. In some cases meat colour is changed, becoming brown if incorrectly frozen. Dehydration occurs during the frozen storage. Dripping during thawing makes the prawn meat more spongy and it loses its original taste. Generally speaking, it is necessary to freeze the prawn at low temperatures as rapidly as possible, and thawing before it is eaten, should be done at moderate speed.

Three important parameters, namely, apparent specific heat, thermal conductivity and ice/water fraction in frozen prawn tissue, are required for the development of a suitable refrigeration system for freezing, thawing and cold storage or transport of frozen or chilled prawn. Prawn muscle contains protein, fat and organic and inorganic salts in solution. As the temperature of the prawn is progressively decreased below its initial freezing point -1.1°C, inorganic salt solution throws out pure crystalline ice, itself getting more concentrated in the process. The freezing point of salt solution thus decreases and progressive freezing takes place following the pattern of freezing curve. Thus the enthalpy (H) of prawn muscle of a given composition is a function of temperature (Q) of prawn. For thawing of frozen prawn most of the latent heat must be supplied in the temperature zone -5° to -1°C, so called, "Thermal arrest" where (dh/dQ) becomes very large. This thermal arrest zone is of interest as biophysical investigations have revealed that at critical freezing rate, when the muscle cools from 0 to -5°C in about 60 minutes, each cell contains just one ice crystal and this is also dependent on the initial temperature of the material. This initial temperature effect is attributed partly to changes in thermal conductivity values of frozen and unfrozen prawn tissues and partly to differential cooling rates. X-ray diffraction examination of the ice component of frozen prawn muscle has revealed the common hexagonal form but with preferred crystallographic orientation of ice crystals. During freezing, proteins become increasingly dehydrated as more tightly bound water is freed and to avoid denaturation, low temperatures in freezing are to be avoided. It has been found (Love, 1963) that freezing of bound water at -78°C (and subsequent storage at -14°C) is irreversible. Riedel (1957) defines bound water to be water unfrozen at very low temperatures. It is actually chemically combined in the muscle at the rate of one molecule of water for every two molecules of amino acid, that is, 0.385 kg water per kg of dry substance.

Since the thermal conductivity of frozen prawn at -30°C is about three and half times that of chilled prawn at 1°C, for same temperature gradient it is easier to freeze than to thaw. However, the electrical conductivity at low frequency of fully thawed prawn is 250 times that of frozen prawn at -28°C, a property utilized in the design of electrical resistance thawing. Enthalpy or apparent specific heat of prawn in the freezing zone is considered to be the algebraic sum of specific heats of salt solution and ice, latent heats of solution of salt, melting of pure and eutectic ice and specific heat of protein gel and unfrozen water. Enthalpy measurements useful in the design of refrigeration equipment have been worked out for cod, haddock, white fish. To work outs ice requirements for chilling and storage of fresh prawn, knowledge of specific heat of prawn above 0°C is necessary. So also the thermal conductivity of prawn above 0°C is important in arriving at optimum thickness of prawn layer for rapid chilling by ice. The bottom of a 10 cm thick layer of prawns iced on the top will be at 10°C even after an interval of three hours due to poor conductivity of chilled fish. Kinetics of biological reaction rates and frequency factors for such reactions have been studied to predict the physical and chemical changes the frozen prawn undergoes during fluctuating or steady low temperature storage. Recent investigations try to link these changes to the unusual dielectric behaviour of frozen prawn. It has been observed that frozen cod muscle exhibit characteristic resistance behaviour of intrinsic semi-conductors like crystalline proteins. Energy of activation of protein denaturation is close to that of charge carriers and there is a possibility that proteins are charge carriers and essential for the chemical process in the frozen state.

Knowledge of variation of density with temperature is required in the calculations of freezing or thawing times and bulk densities for stowage rates. The determination of density is purely theoretical and as anticipated it falls as freezing proceeds, as large portion of moisture in prawn is converted into ice.

Texture of processed prawn, an important quality factor linked with the structural quality of the material, is often defined in terms of hardness, fluidity, elasticity, adhesiveness, chewiness etc. The manner in which these properties manifest themselves, though difficult to analyze, can be expressed in terms of energy or power required to chew which can be measured in an equipment designed to determine its shear strength, elasticity and Poission's ratio to effect that change.

Temperature and Humidity

Temperature and humidity are other important physical parameters in prawn preservation. It is observed that the rate of spoilage of prawn is related to the temperature and at 2.5°C it would be twice as fast as at −1.1°C. The storage life of prawn product at −18°C is reduced by six weeks if it kept for 3 days at −9.5°C prior to storage at −18°C.

Humidity in the cold storages is required to be near saturation, that is, the partial pressure of water vapour in air should be equal to the vapour pressure of ice on the surface of prawn at the storage temperature (-30°C and below) to avoid dehydration (freezer burn) of frozen prawn. Psychometric charts show that at these low temperatures, even a wet bulb depression of 0.5°C is enough to reduce the humidity from saturation by 5 to 6 per cent and the drying potential could be easily imagined. The capacitance resistance hygrometer having anodised aluminium oxide as the dielectric material can be used with certain amount of accuracy at relative humidities above 90 per cent and temperatures upto -15°C. During initial stages of drying too, the drying rate of prawn muscle is controlled by environmental conditions of which humidity is an important parameter.

Canning of Prawns

Canning is a method of preservation where heat treatment is given to food material in air-tight sealed containers, so that all the pathogens will be killed and other micro-organisms may or may not be killed but made inactive. Also the heat treatment inactivates the enzymes, responsible for causing spoilage and thus, contributes to the preservative action. The main advantages in canned products are that they can be stored for years at ambient temperature and are ready-to-eat products. However, cost is the only and major criteria for its production

The flow diagram, for a general preparation of canned prawns is as follows.

Raw material → Washing → Peeling → Deveining → Washing → Size grading → Blanching in boiling 10 percent brine for 4 to 8 minutes → Draining → Fan drying → Weighing for different size cans → Packing → Adding salt etc. → Filling till net weight for that size can → Vacaum sealing → Processing at 10 lb pressure per square inch for 18 minutes → Cooling → Coding → Packing in cartons → Storing at room temperature.

Considerations in Canning Prawns

There are some common defects generally found in canned prawns. The poor appearance and odd flavours of the product are mainly due to the stale raw material and prolonged storage of raw material in ice. Overfilling in no case is desirable. Generally low vacaum is caused due to delay in seaming, defective seams and microbial activity. Reduction in filling media causes non-uniform heat penetration. Balancing is also an important step in the preparation of canned product. An over-balanched material generally takes up moisture and results in over weight and under-balanced material loses moisture and thus, becomes under-weight. Besides neither over-cooking nor under-cooking is desirable. Under-cooking results in microbial survivality while over-cooking destroys the food quality. Sudden cooling is recommended in potable water. Blackening of can interior–and can contents has been a major defect in canned prawns. This is due to formation of iron sulphides from the hydrogen sulphide released from sulphur containing amino acids of the proteins and the metals which contaminate the material during processing.

The pattern and cause of bacterial contamination of canned prawns are the same as those in all heat processed canned foods, namely, under-processing which leaves heat resistant, thermophilic spore-formers in the cans or leakage through seams which causes the entry into the cans of wide variety of organisms of all types from air, water etc. The problem is of special interest to India particularly from the point of view of its export-oriented prawn canning industry.

Canned prawns are processed at high temperature in steam under pressure. This procedure eliminate all bacteria except those having spores of exceptionally high resistance to heat. The presence of only spore forming organisms growing at 37°C and 55°C is generally indicative of under processing. Occurrence of spore-forming as well as non-spore forming organisms in cultures might be due to the cans by passings the retort room without receiving adequate heat processing. The presence of a mixed flora of rods and cocci on microscopic examination and the isolation of these organisms on subculture is indicative of leakage. Spoilage due to leakage may be caused by excessive contamination of the cooling water or damage to the can through rough handling.

In the bacteriologically defective cans (which average about 0.3 per cent of total production) spore formers predominate though present in some cases with a mixed flora. The presence of spore formers in the cooling water point out that the cause of contamination of the cans could be from leakage through seams during cooling process.

A few species of the *Bacillus* type, namely, *Bacillus pantothenticus*, *B. firmus*, *B. brevis*, *B. pumilus* are frequently met within the cans.

Blackening of Canned Prawns

Canned crustaceans, particularly prawns, though packed in special lacquered cans are not free from product blackening and sulphur staining of can interior. Sulphur staining is of two types, one caused by tin sulphide and the other by iron sulphide. Apart from the staining, deposition of iron sulphide at break points of the can interior is also often observed.

In the case of canned shrimp products, outbreaks of iron sulphide discolouration had a profound effect on the export trade, was evidenced by India in the first half of 1967 when more than 84 per cent of the total detention in preshipment inspection was due to the blackening of the material or the can interior.

The blackening canned prawn meat took place by direct interaction between sulphur and copper or iron or by secondary reaction through iron sulphide deposits. In the former case their exist a linear relationship between the intensities of blackening and copper content of the material. Slightly blackened products usually show above 15 ppm of copper (DWB). Meat containing 200 ppm of iron (DWB) or above show a deep brownish discolouration which is distinct from blackening caused by copper. Iron sulphide blackening as against discolouration is characterised by deposition of black spots initially along the vertical seam–joint spreading slowly along the internal seam curvature and imparting black discolouration to the meat in the can.

The main source of copper and iron is the prawn tissue itself, which after canning was found to contain 3 to 12 ppm copper and 12 to 64 ppm of iron (DWB). Water, salt, ice, copper–and iron–based utensils with which the meat come into contacts during the various stages of canning and repeatedly used blanching brine in continuous blanching might also individually or collectively contribute to the mineral contamination. The occurrence of heavy metals associated with water, ice, brine, citric acid and meat at various stages also add to the contamination. Inert type of utensils, such as, stainless steel or plastic vessels or others which form white metallic sulphides, such as, aluminium, may be used for handling the material for canning.

Iron being the most abundant mineral in the world, the chances of contamination are more. However, blackening caused by iron-sulphide is less as its deposition is not favoured under standard conditions of canning and usually the iron content of the meat does not exceed its critical limits. But canners very often drift from the standard technique of processing befitting their need and facility only to face some difficulties. Though the most modern inert-lacquer is used for coating the interior of the tin plated iron-based container, it does not completely prevent the transmission of iron from the body of the can to the contents during storage.

Besides the heavy metals, volatile sulphides are also necessary for the formation of black copper and iron sulphides. The main sources of these sulphides are the sulphur containing amino acids of the tissue and the sulphides sometimes are found in water. The release of sulphides from the tissue depends on;

1. *Quality of the Meat*: The poorer the quality the more is the release of volatile sulphides. During ice storage of the meat, the volatile sulphides show an initial increase followed by gradual fall probably due to loss of solubles by leaching;

2. *Retorting Conditions*: The higher the retorting time and temperature, the more is the release of volatile sulphides due to break down of tissue;

3. *Concentration of Brine in the Can*: The lower the concentration of salt (less than 2 per cent) the more is the release of sulphides; and

4. *pH of the Contents*: The higher the pH of the packed contents, the more is the release of volatile sulphides. It is known that cystine is the main precursor of sulphides in muscle, its breakdown (at 100°C) occurring only in alkaline medium.

Control of acidity in canned prawn is very important in order to maintain its storage life and also to keep it free from undesirable influence of contaminants. Generally, acidity of brine in can is controlled by addition of citric acid in the blanching or filling brine or both. Addition of (upto 0.5 per cent) citric acid in blanching brine and finally packing the blanched meat in simple brine is a common commercial practice.

In India, prawn is usually packed in cans of 301x109, 301x206, 301x307 and 401x411 dimensions. The minimum desirable level of acidity can be maintained so long as the correct ratio of meat and brine is maintained. Prawns packed in the can sizes of 301x109, 301x206 and 301x307 maintain an average ratio of 1.54, but in 401x411 size it is 2.34. So, in order to maintain the same level of acidity in 401x411 cans, it is necessary to add 1.5 times more acid than that required by the other sizes. But canners under standard conditions of canning usually add the same filling brine with or without acid in all cases irrespective of can size. Iron sulphides deposition is thus likely to be more in 401x411 cans.

The species of prawn have no influence on the quality of acid to be added for maintaining the acidity, while the size of the prawn exhibits significant effect, smaller size grades requiring more acid in filling brine compared to bigger size grades. It is therefore necessary to blanch the smaller size grades in lower percentage of acid (0.1 per cent) and then pack in 0.2 to 0.25 per cent acid (in fill brine) rather than the reverse conditions which are commercially practised.

Recommendations

Material used for canning should be fresh and should have minimum level of heavy metal contamination. Only potable water free from copper and iron as far as practicable but not containing more than 0.1 and 0.3 ppm respectively should be used for all requirements in the various stages of canning. Direct use of iron and copper utensils, particularly the latter should be avoided.

Cans should be processed at a lower temperature for longer period (22 minutes at 115°C) rather than at higher temperature for shorter period. (12 minutes at 121°C). Higher temperatures and prolonged retorting should be avoided.

Titratable acidity (0.06 to 0.15 per cent) of the brine in the processed cans should be maintained by the incorporation of specified quality of citric acid in the filling brine depending on the period of pre-processing ice-storage of the prawn.

Quality Standards for Canned Prawns

The taste and smell of the canned prawns should be pleasant. The consistency of the meat should be firm and not tough. The microbiological control has a very important role in maintaining the high quality of finished product in the canning industry. Swelling of cans and spoilage of raw material are due to microbial action. Not only the count of microbes, but also species composition should be determined. Special care must be given for estimation of *Escherichia coli* (indicator organism) *Staphylococcus aureus* (toxicogenic organism). Vacaum inside the can must not be less than 10 cm of Hg. Head space shall be 5 to 7.5 mm and drained weight shall not be less than 65 per cent of water capacity of the can.

Drying of Prawns

Drying in the sun is one of the oldest and most widely practised methods of preservation of prawns. Here the moisture content of the material is lowered to such an extent that the micro-organisms, responsible for bringing about the spoilage are destroyed and the enzymes are denatured and inactivated. Thus the material can be preserved for longer period.

The flow diagrams for the preparation of dried prawns and semi-dried prawns are:

Dried Prawns

Raw material → Washing → Sun drying or mechanical drying → Packing in polythene bag or gunny bag → Storing at room temperature.

Semi-dried Prawn

Raw material → Washing → Blanching by dipping in 4 to 6 percent boiling brine for 2 to 3 minutes → Cooling → Peeling → Immersing in saturated brine for 15 minutes → Drying at 50°C till moisture content attains 40 per cent → Packing in polythene bags → Storing at room temperature.

Considerations for Dried and Semi-dried Prawns

The moisture content of the dried material is very important. Moisture content must be 8 to 10 percent in case of dried prawns, while 40 per cent in semi-dried prawns. Eventhough sundrying is widely practised, artificial dehydration has got some advantages over conventional methods. In the artificial dehydration method, the drying is quicker and more efficient and turns out a much more hygienic product. But disadvantage is the added cost involved in the new method. There are different types of dryers, used for the purpose, namely, hot air dryer, tunnel dryer, rotary dryer, fluidised bed dryer, solar dryer etc. In this method, temperature, relative humidity and wind speed are regulated to a desirable level for obtaining a better product.

Raised platform must be provided for sundrying with crow proof and preferably fly proof enclosures. Good storage rooms with moisture proof floorings and rodent proof walls and ceilings must be constructed for storage until it is marketed. Better and hygienic packing materials like polyethylene bags, gunny bags must be employed. However, retail packing should be done in polyethylene bags.

Drying and salting

Since old times salting and drying has been a popular way of prawn preservation in many countries. The prawn may be shellon and dried in the sun. Where the sunshine is insufficient, it is dried by hot air from a fire.

Dehydration of prawn meat prevents the growth of bacteria and the activity of enzyme as well. Thus the dried prawn can be preserved long. In recent years, however, drying has been carried out, specially for the purpose of improving taste of the product. In this case, the product contains more moisture than simple dried products.

Whole body of shrimp and other shell fish are dried after cooking. The protein contained in prawn meat denaturates through this process for better preservation.

Boiled and frozen shrimp ready for house hold cooking and in processing factories are popular in many countries. Large frozen blocks of shrimp are brought to factory. They are thawed in running water. Head and shell are removed. Washing and sterilizing is done with chlorine water. Small gaps are made in the meat. Bread flour coating is given on cut shrimp. Coated shrimp are packed in small boxes and transferred to freezer. Freezing is done at temperatures 35° to 40°C below zero. Packing in polythene bags are made during the freezing process. The product is shipped to market at 20°C below zero.

Preservation of prawn by salting and drying has been practised since long in developing countries. Yet approach to these practices had been mostly practical and emperical, scientific investigations into the basic mechanisms involved in the process being scanty. During initial stage of constant rate period, the drying rate of prawn muscle is controlled by external conditions and is equal to that from a saturated surface of same shape. The duration of this period is expressed by a relation containing the rate of evaporation per unit area, effective diffusion constant, thickness of the sample and free water content. The subsequent falling rate has two distinct phases. The drying behaviour in the first phase is in accord with the solution of a diffusion equation based on Fick's Law, the effective diffusion coefficient being independent of shrinkage of prawn. The diffusion coefficient in the second phase is about one-fifth of that of the first phase. The transition from the first to second falling rate period appears to be associated with uncovering of the unimolecular layer of water which covers the protein molecule. The process of evaporation and diffusion is characterized by a scheme of energy levels involving the heat of adsorption of unimolecular layer of water, the heat of liquefaction of water and activation energies corresponding to each of the two phases of falling rate periods. The effect of presence of fat is reduction in the effective values of diffusion coefficients in the first and second phases of falling rate period.

Studying the dynamic aspects of salting of fish, it has been observed that salt concentration in prawn muscle and tissue water increases with brine concentration. The equilibrium distribution coefficient based on muscle volume and water content increases at first, passes through a maximum and then decreases with brine concentration. The distribution coefficient is roughly equal to unity and is independent of salt concentration in the brine. The migration of salt in prawn muscle, has been studied deeply by observing other factors, like equivalent conductance and sodium and chloride ion transference number in muscle and water. The diffusion coefficient and equivalent conductance of salt in prawn muscle increase with temperature and time and are in agreement with predicted values. The activation energies for salt diffusion in prawn muscle and in water of infinite dilution are found to be in the region of hydrogen bonding energies.

Diversified Prawn Products

Dry Prawn Pulp is one among the special cured products. Here, the prawns are washed in sea water. Some amount of common salt may be added at times. Then the raw materials are cooked in brine water for sufficient time to obtain desirable characteristic pink colour and curling. The cooked prawns are sundried over a mat and packed inside gunny bags and beaten with a wooden mallet to seperate the shells. Thus the edible pulp is seperated from shell particles and packed in polyethylene bags.

Prawns, peeled and deveined are the raw materials for break prawns. Here they are blanched in 7 to 80 percent boiling brine for 12 to 15 minutes, cooled and kept aside. Then batter is prepared using maida, water, egg and salt. After preparing this batter, the blanched prawns are dipped individually in the batter and coated with bread powder by gently rolling the battered prawn in the powder. This is then fried in edible oil and served. Sometimes the product can also be frozen and preserved for a longtime and fried in oil as and when required.

Prawn flakes is prepared from minced prawn meat. To this minced meat, starch and other ingredients in suitable proportion are mixed well and then steamed followed by cutting into required size. It is then dried in any artificial dryer till moisture content of the product attains at about 5 per cent. It becomes a very goods side dish when fried in edible oil resulting into swelling and becoming crisp in nature. This can be packed in polyethylene bags and stored in room temperature for a long period.

Shrimp extract is a partially hydrolysed prawn protein concentrate powder. The product is very popular in USA, Canada, and U.K. It has got very good taste and at the same time, it contributes to the daily protein requirement of the consumer. Generally shrimp powder of *Acetes* spp is used for the preparation of this product. The shrimp powder after mixing with water hydrolysed partially and the mass is concentrated using agar agar. The product is generally filled in cans, vacaum sealed and processed in a retort at 115.3° C for minutes. The shelf-life of the product is over a year at room temperature.

Small varieties of prawns (P and D) are suitable for prawn pickle preparation. In this preparation ingredients are mixed to the balanced raw material in a fixed proportion. The method of mixing the ingredients consists of heating the oil at 180-200°C in a vessel to which mustard seeds are fried, chilies and lemon pieces are next added and stirred well. Salt, ginger pieces, red chilly, white pepper, turmeric powder, peeled garlic etc are added in the sequence and heated for about 2 minutes. Vinegar is then added and mixed well. After cooling sodium benzoate is added for preservation. The product with its pH around 4.5 is most desirable. The pickle is then packed in bottles and their mouths sealed with acid proof caps. The shelf-life of this pickle is for about one year.

In the prawn processing industry, the important waste materials are its head, shells and tail. All these can be readily converted into prawn manure. This is a highly concentrated nutritious feed supplement containing proteins, minerals, vitamins and some unknown growth factors. Generally they are used along with cattle and poultry feed. It is prepared by drying the waste materials in the sun and grinding to powders.

Chitin is a polysaccharide, found in the hard shells of prawn. Chitosan is derived from chitin by a process of deacetylation. The protein is extracted from the shell with hot dilute sodium hydroxide. The insoluble material is treated with HCL and the product is filtered and seperated solids are air dried for 48 hours and ground to yield chitins. Chitin and chitosan are used for sizing the rayon, synthetic fibres, paper, wood, cellophane, as adhesive, stabilizing and thickening agents. It is also used as chromatographic base, iron exchange resin etc. Chitosan also helps in coagulation of food processing wastes successfully.

Chapter 10
Deep Sea Prawns Landed in Indian Coast

Trawlers of 48-70 feet OAL made of either wooden or steel hulls, coated with fibre glass were used for deep sea prawn fishing. They were powered with 106-140 HP diesel engines and landed their catch in ten fishing harbours of Kerala, namely, Sakthikulangara, Neendakara, Thottapally, Cochin, Munambum, Murikkumpadam, Ponnani, Beypore, Puthiyappa and Mopala Bay during September 2000 to April 2001 which have been quanified as 48675 tonnes. The deep sea trawl units showed almost a double fold increase around 325 numbers, during 2001-2002. Majority of them are more than 50 feet LOA operating with their base at Munambum harbour. Eighty percent of the vessels are above 50 feet LOA and 99 per cent of them are equipped with GPS, Echo sounders. Some of the vessels were found using wireless set also. These vessels operate four seam trawl net having a cod end mesh size of 20 to 22 mm. Total length of the net varied from 130-150 feet. Usually 8 to 9 fishermen go for deep sea fishing trip which lasted 8 to 9 days depending on the season.

The deep sea prawn fishery of Kerala coast was constituted by the following fifteen species; *Parapandalus spinipes* Bate, *Heterocarpus woodmasoni*, Alcock, *Heterocarpus gibbosus* Bate; *Aristeus alcocki* Ramadan; *Penaeopsis jerryi* Perez Farfante; *Plesionika martia* Milne-Edwards, *Metapenaeopsis andamanensis*, Wood-mason; *Solenocera hextii*, Wood-mason; *Acanthephyra sanguinea*, Wood-mason; *Acanthephyra armata*, Milne-Edwards; *Heterocarpus laevigatus* Bate; *Plesionika alcocki* Anderson; *Oplophorus typus* Milne-Edwards; *Parapenaeus investigatoris* Alcock and Anderson and *Plesionika ensis* de Man.

P. spinipes appeared as the most dominant species contributing to 19 per cent and 40 per cent of the total exploited stock during 2000-01 (September to April) and 2001-02 (September to May) respectively, *H. gibbosus* and *H. woodmasoni* for 7980 tonnes (16 per cent) and 7786 tonnes (16 per cent) respectively to the preceding year. However their contribution declined to 10 per cent and 13 per cent respectively in succeeding year. *M. andamanensis* accounted for 14 per cent and 21 per cent respectively of the total catch during 2000-01 and 2001-02. *A. alcocki* formed 12 per cent and 10 per cent of the total deep sea prawn landings in these years. *S. hextii* contributed to 14 per cent with an annual catch of 6640 tonnes. The catch of the species was drastically lowered during the second year (341 tonnes forming 2 per cent of the total landings).

Table 12: Percentage Composition of Various Species to the Total Deep Sea Landings

Species	2000-01	2001-02
P. spinipes	9208t (19 per cent)	7646t (40 per cent)
H. wood masoni	7737t (16 per cent)	2519t (13 per cent)
H. gibbosus	7981t (16 per cent)	1991t (10 per cent)
M. andamanensis	6642t (14 per cent)	4148t (21 per cent)
S. hextii	6640t (14 per cent)	341t (2 per cent)
A. alcocki	5769t (12 per cent)	1799t (9 per cent)
P. jerri	4098t (8 per cent)	34t (0 per cent)
P. martia	600t (1 per cent)	342t (2 per cent)
A. sanguinea	1.4t (0 per cent)	4t (0 per cent)
P. investigatoris	–	493t (3 per cent)

Though the fishing season commenced from September, bulk of the landings was observed during December to March. During 2001-02, the commencement of fishery was observed slightly earlier, from August onwards and continued upto April. In both the years, the peak fishery was observed during December to April, while the fishery was appeared to be very bleak in the months of August and October.

The average catch per hour of deep sea prawns was computed at 12.14 kg during 2000-01, which declined to 7.13 kg during 2001-02. In contrast, the average catch per unit effort showed an increase from 24.8 kg in the preceding year to 31.28 kg in the succeeding year.

Parapandalus spinipes, the "oriental narwal shrimp" was the dominant constituent of the deep sea prawn landed in Kerala. Major portion of the stock was exploited from the depth zones of 190-320 m off Ezhimala in the north and off Thottapally in the south. The catch per hour and the catch per unit effort of *P. spinipes* showed wide fluctuations and the annual average catch per hour showed a drastic decline from 21.74 kg during 2000-01 to 9.18 kg during 2001-02. In the preceding year, the catch per hour varied between 84 kg in September to 0.08 kg in October, while the highest and lowest catch per unit effort could be discernible in November (87.14 kg) and October (0.15 kg) respectively. The fishery of *P. spinipes* was constituted by prawns ranging in length from 51-150 mm in females and 51-160 mm in males. However, specimens below 71 mm and above 121 mm were barely represented in the commercial landings. In *P. spinipes*, the percentage of berried prawns in the landings were very high throughout the year except in July (48.28 per cent). The peak breeding was observed during October to January with highest numbers registered during November (92.15 per cent) followed by December (91.03 per cent). In general, females showed slight dominance over the males in the total landings. About 80 per cent of the population is represented by berried females from September to December with maximum contribution in December.

Heterocarpus gibbosus, appeared as the dominant species in the exploited stock landed mostly from 240 to 380 m depth off Cochin coast. The annual average catch per hour and catch per unit effort of *H. gibbosus* were estimated at 8.00 kg and 20.64 kg respectively. Lowest catch per hour and catch per unit effort were registered in October, while the same was highest in March with 19.81 kg and 36.05 kg respectively. The average catch per hour and catch per unit effort showed a decline during the succeeding year (2001-02), with 5.67kg and 26.47 kg respectively. In males and females of *H. gibbosus*, the length frequency distribution showed an unimodal character with the modes frequently observed at 91-100 mm and 111-120 mm length range. The monthly sex ratio of *H. gibbosus* during the two years indicated that there exists a significant departure from 1:1 ratio during almost all the months due to the dominance of males in the exploited stock. Chi-square analysis of sex ratio showed significant deviation from 1:1 ratio at 5 per cent level. During September and October, the preponderance of juveniles in the exploited stock is note worthy, while the contribution of berried females were comparatively less in these months. However, the berried population could be discernibe from November onwards which attained peak in March.

Highest landings of *Heterocarpus woodmasoni* was recorded from Sakthikulangara. During 2000-01, high catch per hour of 21.91 kg was observed in November, while the catch per unit effort was high in October with 28 kg. In 2001-02, high catch per hour and catch per unit effort of 17.15 and 48.74 kg respectively were observed in September. During the onset of fishery, bulk of the landings were registered from 180-240 m depth, and the sizes were very small and during the peak landing period, majority of the landings was reckoned in between Tuticorin to Quilon at a depth of 260-380 m. More or less the same distribution pattern was noticed by the INP vessels during the exploratory survey. The landings from northern region was negligible. Maximum abundance of the species was recorded in Quilon Bank and 300-400 m depth off Ponnani. In 2000-01, the modal classes of male and female were 91-100 mm and 81-90 mm respectively, followed by 81-90 mm and 91-100 mm respectively. In contrast, during 2001-02 the modal values were represented by two length groups of 91-100 mm and 101-110 mm both in male and female population followed by 81-90 mm in both the sexes. The Chi-square value showed a significant deviation from 1:1 ratio during September to December in both the years due to the predominance of females in the catch. A further spurt in the values was observed during May to August due to the abundance of females. In *H. woodmasoni*, the peak occurrence of berried females was observed during December to February, when on an average 88.7 per cent of females were found to carry eggs attached to the pleopodal setae. The second major peak was observed in October, when 81.4 per cent of the total female population were found to carry eggs. From March, onwards, there was a steady decline of egg carrying females to half and the lowest number was registered during July with only 22.75 per cent berried females in the total female population.

Metapenaeopsis andamanensis, the rice velvet shrimp, is one of the common penaeid present in appreciable quantities during December to April with peak landings in February. This species is often found landed along with *P. spinipes* from relatively shallower grounds from 200-280 m, off Cochin coast. The fishery of *M. andamanensis* was constituted by individuals in the range of 32 to 148 mm in females and 39 to 141 mm in males in the commercial landings. The male to female ratio of *M. andamanensis* in the exploited stock was 1:2.44 during 2000-01 and 1:1.87 in 2001-02, thus showing the preponderance of females in the population. The month wise analysis indicated a significant deviation from 1:1 ratio in September to April in both the years owing to the predominance of females.

Aristeus alcocki, "the Arabian Red Shrimp, locally known as "Red ring" is the most valuable deep sea prawn due to its demand in the export market. The annual average catch per hour of *A. alcocki* was computed at 8.64 kg during 2000-01, which declined to 3.22 kg in the succeeding year, whereas the catch per unit effort showed an increase from 27.96 kg in the first year to 37.77 kg in the second year. The highest catch per hour of *A. alcocki* during 2000-01 and 2001-02 was recorded in February and January respectively, while the lowest value was in September. This species was not observed in the landings from the vessels operated at depth ranges less than 300 m, the maximum catch been recorded from 380-550 m depth and were mostly observed from two fishing grounds, off Ezhimala in the north and off Quilon in the south. The existence of these two prominent fishing grounds of *A. alcocki* along Kerala coast have already been reported during the exploratory surveys conducted during 1968. This species represented the fishery throughout the season in varying proportions. In females, 121-130 mm appeared as the modal size class followed by 101-110 mm and 131-140 mm. Males found to be smaller than females in the exploited stock. Males *A. alcocki* of 81-90 mm formed as the modal class during July, August, while 101-110 mm length groups frequently represented in the catches during rest of the months. The monthly sex ratio showed a significant departure from the hypothetical ratio 1 : 1 in all months except in July due to the predominance of females in September and December; while males predominated in the rests of the months.

Plesionika martia, "the Golden Shrimp", is comparatively smaller pandalid prawn landed in minor quantities from November to

February in the depth range of 180-330 m. The total length varied from 65 to 102 mm in males and 73 to 102 mm in females. However, the size groups of 81-86 mm in males and 90-98 mm in females showed predominance. The landings could be observed only from November to February with peak in December. The highest catch per effort was observed in January with 22 kg per hour, while it was lowest in November (0.94 kg/hour).

Solenocera hextii, a larger prawn often known as "the Deep sea Mud Shrimp" was represented in the catch throughout the fishing season except in October, mainly exploited off Cochin from 130-132 m depth. Highest quantity was landed in December, February and March, and the females out numbered the males in the landings and almost all mature females were in the impregnated conditions. Size varied from 89 mm to 140 mm in males and 130 mm to 149 mm in females, but the size groups having preponderance of 110 mm to 130 mm of the former and 129 mm to 140 mm of the later in the landings was quite discernible.

Penaeopsis jerryi, the "Gondwana Shrimp" is a widely distributed small penaeid represented the landings from the depth of 251-300 m. Peak period of abundance was during January to March. Size range of 58 mm to 108 mm in males and 67 mm to 114 mm in females contributed to the fishery. The modal sizes between 76 mm to 89 mm dominated in the catches in both sexes.

Acanthephyra sanguinea, a deep red coloured prawn, often mistaken as *A. alcocki* by the fishermen is landed in minor quantities and represented the fishery in April at 320 m depth off Cochin. Size range of males was 88 to 100 mm and 102 to 118 mm in females. The impregnated females outnumbered the males in the total landings.

The deep sea prawn landings showed wide fluctuations along Kerala coasts during September 2000 to April 2002. The monthly distribution of catch per effort indicates that December is the peak month of abundance followed by February and March. The exploratory surveys during 1985 revealed the existence of 14 species, though fifteen species could be recorded in the commercial landings in varying proportions. Among the 15 species of deep sea prawns landed at various harbours of Kerala, the commercial fishery was mostly confined on five species, namely, *A. alcocki, H. woodmasoni, H. gibbosus, P. spinipes* and *M. andamanensis* in their order of preference. By virtue of excellent demand for export, *A. alcocki* was exclusively procured by the seafood processing plants. As a result, with the onset of fishery, majority of fishermen started selective harvesting of this species of high value by embarking in the grounds known for their predominance. *P. spinipes* appeared as the most dominant species in the total deep sea prawn landings contributing 19 per cent and 40 per cent respectively during 2000-01 and 2001-02. *H. gibbosus* and *H. woodmasoni* contributed 16 per cent each during the first year and occupied second position in the landings. However, their contribution declined to 10 per cent and 13 per cent respectively, shifting to third position during the second year.

Due to the ever increasing demand for prawns from the processing industry, deep sea trawl units engaged in trawl fishery showed almost a twofold increase during 2001-02, compared to 1999-2000 and consequently there was an exponential increase in the fishing effort within a short period of two years. By 2003, nearly 300 shrimp trawlers have been converted for deep sea operations by fitting GPS and echo sounders. Besides more than a dozen of new fishing crafts were commissioned in and around Munambam harbour exclusively targeting for deep sea prawn fishery. Observations on the spawing biology of deep sea prawns reveal that the peak spawning is more or less synchronizing with peak fishing season. The data of month wise and lengthwise sex ratio analysis brought out the preponderance of females over males in a number of commercially important species, such as, *P. spinipes, H. woodmasoni, M. andamanensis* and *A. alcocki*. This skewness in the sex ratio by females would suggest the possibility of differential migration of male population from the fishing ground and this can be postulated as one of the reasons for the stock depletion of deep sea prawns. Percentage of berried pandalid prawns were found to be very high during December to March, in the range of 71.33 to 91.25 per cent and a decline of the fishery registered during the second year can well be attributed to the indiscriminate exploitation of berried females by the commercial fishing units.

Protection of the breeding stock, prevention of growth of over fishing, annual closure of the fishery during south west monsoon and imposition of restricted fishing season together with strict regulation of the units put under operation are some of the options for the sustenance of the stock. But the ever increasing demand for export purpose, may further aggravate the fishing pressure even at higher depths and hence there is every possibility of stock depletion in near future. The indiscriminate exploitation of berried population of deep sea prawns may lead to poor recruitment over fishing. The stock of deep sea prawns would be in a dangerous situation in near future unless otherwise the fishery is regulated at optimal levels giving due attention to maximum sustainable yield, stock-recruitment relationship and growth rate of individual species.

Chapter 11

Marine Aquaculture: Culture of Marine Prawns

Global marine aquaculture production has expanded over the years from 5.5 million tonnes in 2002 to reach a record high of 7.1 million tonnes in 2006.

There has been overfishing of marine fishery resources round the globe, resulting in the exploitation of 75 per cent of the stocks or upto their biological limits. To give relief to the marine capture fisheries sector from the strain it now suffers, it is estimated that a five fold increase in global aquaculture production is needed within the next five decades to maintain the current aquatic food consumption levels. In 2003, the per capita global fish supply including China was 16.3 kg/person and 13.3 kg/ excluding China. On a modest rate, a 30 per cent increase in per capita consumption to 25 kg by 2025 and to 30-40 kg by 2050 is estimated. To cover up the widening gap between global fish demand and supply, the global aquaculture production of 52 million tonnes recorded in 2005 would need to be increased to 61 million tonnes by 2010, and further on to 120 million tonnes by 2025, and 210 million tonnes by 2050.

Land based aquaculture operations use land and water not only for fresh but also brackish and/or marine water. On a global scale, water resources becoming scarce and expensive. About 41 per cent of the world population today lives in water stressed river basins. In 2050, 70 per cent of the world population will face water shortage.

Global aquaculture productions increased from 0.64 million tonnes in 1950 to 51.39 million tonnes (valued at US $ 59.99 billion) in 2002. Asia accounted for the bulk of aquaculture production (95 per cent). China accounted for 71 per cent of global farmed fish production while India's share was only 4 per cent.

Shrimp aquaculture expanded rapidly in South-East Asia until mid 1990s. Until then land clearing and construction of shrimp ponds in the region took place. This land clearance was the main cause for the loss of mangrove forests and other forms of coastal degradation. World shrimp production has leveled off in recent years, as many aquaculture farms have either collapsed or have been experiencing declining yields. Lack of sustainability is due to the many factors, such as, poor management and low quality of seeds stocked and over exploitation of resource. Knowledge of ecological factors controlling pond production is also crucial for more effective management and sustainability of prawn farming. Intensification of shrimp farming not only resulted in the concomitant loss and damage to coastal habitats, but also the degradation has been especially severe for mangrove forests.

Farmed shrimp contributes a major share in the total shrimp production and 88 per cent of farmed shrimp is accounted from Asia. Prawn farming constituted one of the phenomenal commercial success stories of the last two decades, with annual growth rate of 20-30 per cent in contrast to stable 2-3 per cent increase in the capture fisheries.

The sudden spurt of shrimp production in leading Asian countries (Philippines, Thailand, Taiwan and China) had accelerated to pollution and disease problems. Other ecological effects of shrimp farming are the loss of mangrove ecosystem, nutrient enrichment, eutrophication in coastal waters, longevity of chemicals and toxicity to non-target species, development of antibiotic resistance, introduction of exotics.

Shrimp farming provides livelihood to about, 120000 farmers in the country. Shrimp aquaculture in India boomed during 1990-95. However, the activity collapsed during 1995-96 mainly due to outbreak of white spot disease.

Excessive accumulation of toxic inorganic nitrogen is always posing major threat to pond ecology, thus not only deteriorating pond environment, but also the environment of the surrounding aquatic ecosystems. Further, nitrogen plays a key role in aquaculture, serving a dual function both as a nutrient and toxicant.

High quality protein is an essential ingredient for the shrimp to attain faster growth and achieve the harvestable size within a short period. Feeding of shrimp under farming is very expensive, as shrimp requires 45 per cent quality protein in the supplementary feed, thus making the shrimp feed very costly, and accounting for more than 55 per cent of the recurring cost of farming operations. On the other hand, 60-70 per cent of the Nitrogen added to the shrimp farms remain unconsumed and wasted (the harvestable part of Nitrogen added to shrimp farms is less than 30 per cent and the remaining 70 per cent is lost as waste). The Nitrogen, thus wasted is subsequently converted into nitrite nitrogen, TAN etc, which are highly toxic to shrimps. Protein rich feed is a major source of ammoniacal nitrogen in the pond. Besides shrimps' excretion of the Ammonium N further aggravates the toxic levels in the pond due to accumulation of these inorganic products in the pond bottom. Shrimps exposed to such inorganic products are subjected to stress and strain, causing the high possibility of out break of shrimp diseases. Accumulation of such toxic materials in the pond bottom will also aggravate the level of pollution not only inside the pond but also in the external environment through the discharge of pond effluent water having high concentration of inorganic nitrogenous products such as ammonia, nitrite-N etc.

Prawn Farming in Southeast Asia

There has been a phenomenal rise in the production of farmed shrimp over the past 2 to 3 decades, made possible by conversion of about one million hectares of coastal land into shrimp ponds coupled with technical breakthroughs in prawn breeding and larval rearing; in shrimp nutrition and formulation of dry and complete feeds; and in the engineering of farm water supply and aeration systems.

There has been a long history of trapping and rearing wild shrimp in several parts of the world, and particularly in coastal lagoons and mangrove swamps in south east and south Asia. Coastal ponds or "tambaks" for rearing milk fish and shrimp probably date back 600 years in Indonesia, while the earliest shrimp ponds in the proper sense may have been those introduced at the turn of twentieth century by Chinese immigrants. Shrimp productions in ponds remained at a low level (average 200-300 kg/ha) until the 1970's when a breakthrough in the artificial breeding and larval rearing of tiger prawn, *Penaeus monodon* was achieved in Taiwan. Until the early 1980's, nearly all marine prawn farming was based on converting mangroves into ponds and the use of tidal flow for water exchange.

Initial Spurts in Prawn Farming

Many of the shrimp farms in south east Asia were constructed in former mangrove areas and suffered from acid sulphate conditions. This is much less of a problem in modern farms, which can pump sea water to neutralise pond acidity. In Thailand, pond dykes are heavily limed and then surfaced with laterite to reduce acid run-off. A potential solution to the problem of physically or chemically unsuitable soils is to line ponds with a plastic material. Although expensive, plastic linings also enable better control of the pond environment, facilitate waste removal, population estimation and harvesting, and greatly reduce pond preparation time. Although a new concept, plastic liners are used successfully in commercial project in Indonesia. There has been a strong trend to site modern prawn farms (encouraged by Government pressure to conserve the remaining mangroves) above the intertidal zone in low-grade agricultural land and to rely on a pumped water supply. Examples include the use of coastal paddy land in Thailand and former sugar plantations in Negros (Philippines).

Although every site is different, many intensive prawn farm use a square, one hectare pond as the standard production unit. From one metre previously, these ponds usually have a water depth of 1.6 to 1.8 m. and great emphasis is made on water exchange to maintain good growing conditions for the stock. There has also been a trend towards directly pumping sea water, although many engineering problems have arisen in trying to construct reliable, cost effective seawater intake systems. Most large projects depend on a reservoir to store sufficient water to enable upto 30 per cent water exchange per pond each day. The reservoir also acts as a sediment trap and it can also be used for mixing freshwater (if available) for greater salinity control. Artificial aeration is essential for intensive culture (approximately one aerator per tonne of shrimp), but has also been proven to increase the growth of prawn even at low stocking densities. A recent innovation has been to orient the aerators carefully in a circular manner in the ponds to provide not only aeration, but also a circular water flow to help accumulate wastes into the centre of the ponds.

Prawn seed supply is supported by more than 3000 hatcheries in Asia, both large-scale and "back yard", but good seed quality and good handling are crucial to successful pond production. There is great scope to improve post-larval quality, particularly the apparent variation between batches and from hatchery to hatchery. Seed quality problems are undoubtedly a contributory factor to the wide range of survival seen between ponds and some farmers purposely overstock on the expectation of 40-50 per cent mortality in the early part of the cycle. Recently, in addition to the standard physical checks on post-larval quality, attempts have been made to use physiological stress tests to identify good quality seed, by exposing post-larvae to a sudden change of salinity. Microscopic and biochemical checks are also being considered.

Although knowledge of prawn nutrition still lags well behind that for livestock and poultry, the south east Asian prawn industry is well supplied with high quality formulated shrimp feeds produced by large international companies as well as smaller national ones (President Enterprises corp, The Hanaqua Group, CP Group, Gold coin). While better diet formulations will continue to evolve through research, prawn farmers also now pay much greater attention to develop better feeding practices to improve conversion efficiencies and reduce waste. Apart from the financial loss, food wastage is also recognized as a major source of water pollution. Great care are given to calculating feeding rates, based on the inspection of good consumption from feeding trays, plus estimates of population size. Young prawn are usually fed 5 to 6 times per day and great care is taken to the way food is distributed into the pond.

Pond management covers the whole range of husbandry measures followed during a grow out cycle. On intensive farms efforts are being made to develop better soil preparation, pond fertilization and other chemical or biological measures to control plankton production, particularly the commonly faced problem of plankton "crashes". Better waste control is also a major concern to reduce the risk of "self-pollution" due to effluents contaminating the source of intake water over a period of time. As mentioned above, good results have been obtained by aerators to direct the movement of wastes into the centre of the ponds, from where they can be drained or siphoned out. There is also some experimental work underway on possible biological methods to reduce the amount of organic wastes in farm effluent water before it is discharged back to the sea, namely, use of mussels or tilapias.

Decline Dues to Disease Problems

A large number of different disease conditions have been reported from prawn farms and hatcheries in south east Asia, including diseases caused by viruses, bacteria, fungi, protozoans and other parasites. Viral diseases, particularly MBV (Monodon Bacculo Virus), has had a wide–spread and damaging effect on the industry. The extreme example is Taiwan which achieved a dramatic rise in tiger prawn production year by year in 1980s, reaching a peak of 80000 mt in 1987. Output then crashed to 25000 mt in 1988 as 40 to 80 percent of the farms became infected with MBV. Initially the industry relied on the use of drugs to control diseases, especially in hatcheries affecting the larvae. Uncontrolled use of antibiotics led, in some cases, to resistant strains of bacteria, or larvae with little disease resistance in normal conditions. There is now a good base of expertise on prawn diseases and realisation that most disease problems occur when prawn become stressed by poor environmental conditions, by overcrowding or by inadequate nutrition. Much more attention is consequently paid to the prevention of diseases by maintaining the prawn in a non-stressful environment. The contribution of overcrowding to disease risks has prompted some experts to recommend a maximum stocking density of 25-30 pls/m^2 for tiger prawn culture.

Culture of Marine Prawn

The reason that species of the Penaeid prawn family have become the successful object of culture fisheries is that they possess three important biological qualities that make them highly suitable for culture. They are (1) they produce an extremely large number of off spring, (2) they are a species with one year life cycle and their reproduction cycle is very short and (3) they grow to a substantial middle-sized prawn in a very short period of time.

Modern marine culture aims to realize control of all stages of production, from collecting the seeds in a water tank to the raising of mature adults and then saving of eggs to seed the next generation. This is referred to as complete or "through culture". In *Penaeus japonicus* culture today, all the steps of the culture process have been perfected, with the exception of raising and keeping parent prawn (Spawners) to parent the next culture generation.

The technological objectives of marine prawn culture can be summarized on the following three points:

1. To improve the survival rate of individuals and achieve a higher concentration of feeding culture than natural seeds, thereby vastly increasing marine production;

2. To maintain a favourable life environment throughout the life cycle of the species involved, thus achieving high feeding efficiency; and

3. To raise healthy individuals with a taste equal to or better than naturally raised ones.

Seed Production

The berried females used for spawning are always naturally grown prawns. Suitable female prawns have been selected, one by one, examining the belly to make sure they have reached full period and the eggs are ready for laying.

The spawning tank, a concrete one, is filled with clean water that has been filtered through sand or charcoal and the water is kept on aeration. In a 100 cubic meter tank 20 to 50 parent prawns are put in. The most suitable water temperature is between 26° and 28°C. If the water temperature is too low, eggs laying does not occur, in which case the water must be heated. If the parent prawns have reached full period, they will usually lay their eggs the first night after being put into the tank. If they don't lay on the first night, they will lay on the second. The parent prawns are kept in the tank either one day and one night or two days and two nights, after which they are removed. If the egg laying is allowed to continue for more than two days there will be considerable size difference between the youngs, making it difficult to properly control their growth in a single tank.

In the past, after the parents had laid their eggs not been removed from the tank, the eggs were raised in the same tank. But in recent years, the practice of removal of fertilized eggs to another tank have been adopted to prevent possible contamination from any diseases which the parents might have been carrying.

The cultivation tanks to be used are first cleaned thoroughly with a high-pressure washing device to get rid of any attached organisms and then dried completely before filling with filtered water. The hatching of fertilized eggs begins about twelve or thirteen hours after being released by the parent. The ideal water conditions for the cultivation tank are believed to be a water temperature of 26° and 28° C, a salinity of 33-35 ppt and pH of 7.0 to 8.0. During cold weather, a boiler is used to heat the water, while in hot weather a sun screen is hung over the tank to prevent temperature rise. The newly hatched prawns proceed to grow through their nauplius, zoea, mysis and post larva stages. During the nauplius stage no feeding is necessary, but as the young enter the zoea stage they begin to feed actively. As the young progress through their growth stages, a progression of feeds is used beginning with phytoplankton, small zooplankton (*Brachionus*) and then middle-sized zooplankton (*Artemia*), small shrimps and composite feeds. In the early part of the post larva stage the young prawns are free-swimming organisms, but as they enter the later part (about PL 30), they begin to display a habit of crawling on the sand. When they reach PL 30-35 (about 40-45 days after hatching) they should be removed from the tank and transferred to a cultivation pond with a sand bottom. Although the number of young kept can vary with the size of the cultivation pond, in general there should be about 20,000 individuals per cubic meter of water at the final stage before transfer. The original number of eggs with which the pond should be stocked is calculated from this number, based on a 60 per cent survival rate. Under the optimal conditions the survival rate from the nauplius stage to PL–30 is as high as 80-90 per cent, but is usually in the range of 50–60 per cent.

Setting Up of Prawn Hatchery

The following criteria should be fulfilled while considering for site selection of prawn hatchery.

A. Land Availability

For a production level of 25 million of PL 20 per year or 1.25 million PL 20 per run of 15 day i.e. 2 run per month and 20 runs a year about 20 hectare of land will be required.

B. Hydrology

1. Salinity of water should be more than 30 ppt and the saline water sources should be far away from rivers. There should be freshwater resources also.

2. The water should be clean and unpolluted, away from industrial and agricultural pollution e.g. factories and paddy lands. The water should be clean with no turbidity and no heavy metals if inshore well is used.

C. Topography

The proposed site should be less than 50 m from intertidal area if water intake is to be made by pipe. The land should be situated 2-3 metres above high tide level and should have sandy or rocky-sandy substratum. The site should be protected from cyclones, storms and other wind and wave action.

D. Source of Broodstock and Spawners

Wild spawners and brood stock should be available nearby from offshore trawlers and from the operators of traditional fishing gear, namely, bottom set gill nets. Brood stock should also be raised in farm ponds.

E. Infrastructure

As regards infrastructure facilities, availability of electricity, accessibility by roads, proximity to farms and other markets for fry and freshwater supply for hatchery and domestic needs are essential.

F. Social Services

The site should be nearer to town for amenities like housing, market, educational and medical services for hatchery staff. The security for the hatchery is to be ensured.

Operation and Management

The stages of development in prawn hatchery consists of fertilized eggs, nauplius–I to VI, protozoea–I to III, mysis–I to III and post larva. Their mode of living include pelagic and continuous swimming in nauplius to protozoea stage, pelagic and swim outbursts in mysis and continuous alternate sedentary and swimming in post-larval stage. The larvae feed throughout the day.

Their food requirements are; in early stages (Z_1–Z_3) they can survive on phytoplankton alone. But subsequently in later stages, they survive on phytoplankton, zooplankton and prepared diet.

Food Requirements

The food required for hatchery production of shrimp seeds are given below.

Phytoplankton

Chaetoceros or *Skeletonema* or *Tetraselmis*

Zooplankton

Artemia nauplii, *Brachionus* (copopods), Cladocerans

Artificial Feed

Microparticulate or Microcapsulated or Prepared diet. The criteria for selection of food for prawn larvae are:

1. Food must be perceived by the larvae.
2. Food must be such that it can be accommodated by the mouth of the larvae.
3. Food can be easily produced in large quantities.
4. Food can be digested by the larvae.
5. Food must satisfies the nutritional requirements of the larvae.

Rearing Facilities

Laboratory facilities for algal starters and algal stock should preferably be air-conditioned. There should be incubators, preferably, cylindro-conical with variable volume. The larval food tanks should either be circular or rectangular or square, but with semi-circular corners and sloping towards centre and outlet. The usual volume ratio of diatom (algal) tanks to larval rearing tanks is 1 : 5. Rotifer rearing tanks are not necessary. Hatching facilities for *Artemia* cysts should preferably be cylindro-conical.

Larval tanks should be either circular or rectangular or square, but with semi-circular corners and sloping towards centre and outlet. The size of the tank is variable, but 10-30 tons are easier to manage.

Nursery tanks should preferably be a part of hatchery system. With a mean survival of 30 per cent from nauplius to early post larva stage, the volume ratio of larval rearing tanks to nursery tanks should be 1:3.

Feeding Programme and Water Management

Diatoms are given as food as early as the protozoeal stage. Zooplankton are given at the beginning of mysis stage. Formulated diets are given as early as the protozeal stage.

Algal density may be maintained more or less the same throughout the rearing period. *Artemia* density is kept lower in the earlier stages, but maintained higher in the older stages.

Amount of formulated diet varies with the number of different kinds of food given, but generally, the amount of formulated diet given increases with the age of larvae or fry.

30 to 50 per cent of the water should be changed daily starting from late Z_1 stage.

Infrastructure Requirements

A. Maturation Area

1. 9 maturation tanks, each of 4 m diameter x 1.2 m depth with 12 cubic metre capacity. The tank should be circular, made up of concrete housed in building with dark roof and walls.

2. 40 spawning tanks, each of 1m diameter x 0.35 m depth with 0.2 cubic metre capacity. The tank should be conical in shape, made up of fibreglass with wooden stand, housed in building with dark roof and walls.

3. 7 broodstock holding tanks, each of 2 m x 1m x 0.6m with 1 cubic metre capacity. The tank should be oval in shape, housed in building with dark roof and walls.

B. Larval Rearing area

1. 20 Larval tanks, each of 5 m x 1m x 1m with 5 cubic metre capacity. The tank should be rectangular made up of concrete, housed in building with 30 per cent skylight and walls.

2. 20 natural food tanks, each of 2m x 1m x 0.6m with 1 cubic metre capacity. The tank should be oval, made up of fibreglass, housed in building with skylight and open.

3. 20 training tanks, each of 1m diameter x 0.35m depth with 0.2 cubic metre capacity. The tank should be conical in shape, made of fibre glass with wooden stand and housed in building with 30 per cent skylight and walls.

C. Post Larval Area

50 post larval tanks, each of 5m x 2m x 1m with 10 cubic metre capacity. The tank should be rectangular made up of concrete and housed in open building with skylight.

Equipments and Other Requirements

A. Electrical

Standby generator

B. Air Supply

Blowers, motor and accessories

C. Water Supply

1. Freshwater well and pump
2. Sea water inshore well or bottom intake pipe
3. Sea water pump
4. Sea water reservoir
5. Sand filter
6. Inline filters
7. Submersible pumps.

D. Laboratory Equipments

1. Pocket refractometer, 0-100 ppt.
2. Dissolve oxygen meter
3. Portable pH meter
4. Compound binocular microscope
5. Stereoscopic (or Dissecting) microscope
6. Neubauer hemacytometer
7. Sedgwick rafter counting cell

8. Balance–0 to 5 kg to weigh chemicals and fertilizers
9. Toploading balance–0 to 500 g to weigh individual prawns.

E. Natural Food

1. Refrigerator to store algal cultures
2. Air–conditioning unit

F. Feed Preparation

Freezer to store broodstock feed.

G. Vehicle

4 wheel drive to transport specimens, broodstock and fry.

Intermediate Nursing

The post larval stage of prawns that have been raised in the cultivation tanks are transferred to an intermediate nursing pond for a 30 to 50 day nursing period in which they are raised to a body length of 4 cm (0.5–1.0 grams). After this they are released into "grow out ponds." For the intermediate nursing, a seperate small-area pond is usually used, but sometimes a small part of the grow-out pond is simply sectioned off with nets to serve as the nursing stage pond.

In the nursing pond a concentration of 70-80 young per square meter of water, or, in the case of high intensity culture, 130-150 per square meter is maintained. The feed used in this stage is all composite feeds.

The intermediate nursing stage has two aims, one is to prevent the young from being eaten and the other is to control the population intensity.

The youngs are nursed and observed closely in a small, confined area to prevent them being eaten by other fish. Also this stage allows the development of sufficient mobility and defensive capacity before the young are released into the grow-out pond. The number of young is also counted carefully at this stage before release into the grow-out pond. This is an essential step in the culture process that enables the control of the culture intensity and aids in determining the amount of feed to be given.

Grow-out Ponds

There are three basic types for construction of grow-out ponds for *Penaeus japonicus* culture.

In partial embankment type, the areas with a large tidal drop is used and an embankment is built up to a height half way between the high and low tide water levels and nets are hung along the top of it.

Figure 45: Partial-Embankment Type Grow Out Pond

In double-bottomed round tank type, a round tank is built on land near the coast and water pipes and pumps are used to supply and drain the water of the tanks.

In full embankment type, an embankment with a water gate is built in the sea or on land adjacent to the sea.

All these three types have evolved their unique features in response to aspects of the natural life patterns or the natural productivity of the area. These differences are most evident in the method of water exchange and the use of diatoms (Table 13).

Figure 46: Full-Embankment Type Grow Out Pond

Figure 47: A 100-ton Type Cultivation Tank (10 × 5 × 2m)

Table 13

	Water Exchange in the Pond	Growth Of Diatoms	Pond Cleaning
Partial embankment type.	Water exchange is achieved by means of the overflow of water over the embankment with the rise and fall of the tides. Each day about one-third of the water is exchanged. A disadvantage exists in the lack of water exchange at low tide.	Natural condition	After harvesting water level is lowered in the pond and new sand put in
Full embankment type	Water exchange is achieved by opening and closing a water gate. At the same time new water is pumped into the pond.	Diatom growth controlled by watching the water colour. The supply of oxygen to the water is the main object	After harvesting the pond is completely drained and sun dried for several days and then sanitized. The bottom sand is also replaced.
Double-bottomed round tank type	A forced exchange of 100-300 per cent of the water is achieved by pump every day.	Diatoms do not grow here	After harvesting the pond is completely drained and sun-dried for several days and then sanitized. The bottom sand is also replaced.

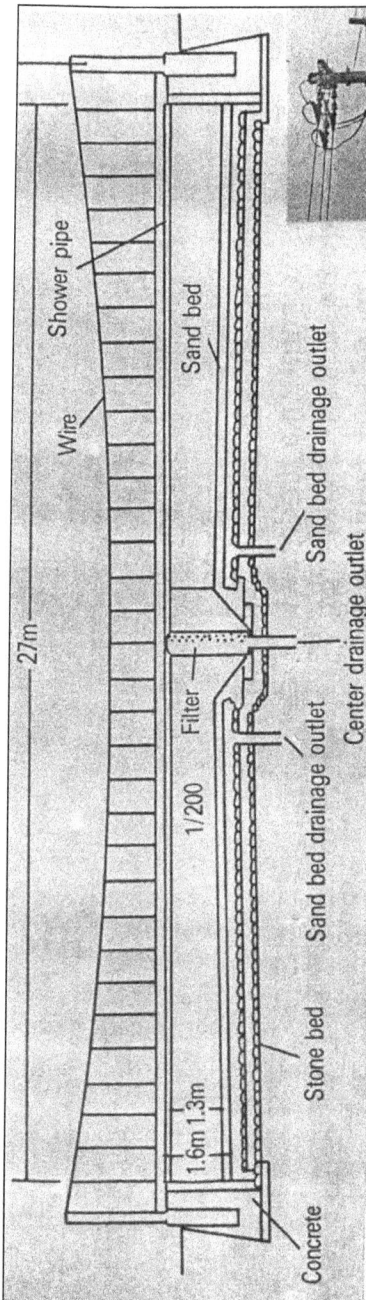

Figure 48: Cross Section of Double Bottomed Round Tank

Figure 49: Double Bottomed Round Tank

The effects of diatom on growth include; (a) increasing the amount of oxygen available in the water through photosynthesis; (b) the brown water colour that diatoms create decreases the amount of sunlight reaching the bottom and thus inhibits the growth of green sea weeds; and to serve as nourishment for the small zooplankton that the prawn fry feed on. However, compared to partial-embankment type ponds, the degree to which the cultivation water is controlled in full embankment type operations is increasing and the role of diatoms as a nutrient is becoming less important. In the case of double-bottomed round tank culture operations, water exchange is so complete that none of the effects achieved by diatoms are relevent any more. The aim of increasing artificial control of the grow-out pond environment is of course, to increase productivity. This is done by (a) increasing the number of young the pond is stocked with, (b) replacing natural feeds with composite feeds, (c) increasing control over the pond water to achieve a more balanced life environment and (d) increasing the harvest for a given area of grow-out pond surface.

For coastal aquaculture of prawns, it is considered important to pay careful attention to the following points, while choosing a site for construction of a prawn farm.

The Topography and Tidal Regime of the Area

The most important consideration is the elevation of the site in relation to the tidal amplitude at the site. The tidal amplitude at the site should be around 1.5 m and the pond bottom should be a little above the mean low water neap tide level so that the pond could be drained completely at any low tide and could be filled with water, if necessary, at every high tide. For proper pond management the ponds should be capable of being drained at least once a year and allowed to dry in the sun.

Ideal location for a farm is a tidal mud flat which is uncovered during low tide and is away from the main flow of the river or creek. It is essential that the farm is not affected by flood waters during the rainy season.

Soil Characteristics

The soil has a direct bearing on the productivity of the ponds. In brackish water ponds benthic productivity is more important than the production of plankton. A clayey soil rich in organic matter encourages the growth of benthic blue alage which associated with microorganisms form the main food of most of the brackish water animals. The microorganisms responsible for nitrogen fixation and mineralisation of organic matter also thrive well on muddy bottom. The pH and carbon content of the soil are also important from the point of view of productivity and recycling of nutrients.

Soil rich in clay is impermeable to water and can be used to form a firm, leak proof bund which is not easily eroded by wave of the tidal action.

Water Characteristics

The most important factor is salinity. Although the juvenile penaeid prawns are euryhaline, the salinity range should normally be 10-35 ppt. Lower and higher salinities retard the growth of the prawns. Low salinity also makes them less resistant to parasitic diseases.

The optimum pH is around 8.0. pH lower than 7.0 or higher than 9.0 is detrimental to the health of the prawns. The methyl orange alkalinity should be 1.2 ml, to guard against large fluctuations in pH caused by the metabolic activity of the plants and animals in the pond. The dissolved oxygen content should not be lower than 3.5 ml/litre. Although prawns can tolerate lesser oxygen concentration, their growth is affected. The turbidity of the water should not be high as it would reduce light penetration and effect phytoplankton production

Seed Resources

If the ponds have to be stocked with prawn seed from natural sources, the abundance of the prawn seed resources in the neighbourhood should be assessed. It is important that the prawn seed available should be those of the faster growing, large species, such as, *Penaeus monodon P. indicus P. merguiensis* and *P. semisulcatus* which fetch a high price.

Flora and Fauna

The area should be free from rooted or floating angiosperms and a dense growth of filamentous green algae is also not encouraging.

Pollution

The site selected should not come under the influence of industrial effluents, city sewage out–falls or insecticide affected agriculture drainage. They would not only affect the growth of the cultured prawns but also render the farm products unacceptable to consumers.

Accessibility

The marketing facilities available for disposing the catch and the logistics of procuring the inputs necessary for constructing and operating the ponds should also be considered.

Relevant informations on legal regulations should be gathered from Revenue and Forest Deptt. to cross the hurdles. For managing the farm efficiently details regarding the seasonal availability of farm labour, local wages paid for skilled and unskilled labours local traditions and customs etc should be collected.

Construction of Prawn Farm

A viable farm unit could have a water spread area of about 10 ha. A well laid out farm should have nursery ponds, stocking ponds and feeder canals. The lay-out of the ponds should be decided, taking into consideration the local topographical conditions. Each stocking pond may have direct and seperate access to the feeder canal, or all the stocking ponds may open into a common catching or "division" pond which has the main sluice gate.

The entire farm should be set apart from the estuary or creek by strong embankments which should stand one metre above the highest flood level. It should be sufficiently broad to withstand the dynamic force of the tides and pressures created by the difference in water levels on both sides of the bunds.

Normally the top of the bund should be as broad as it is high and the slope varies from 1 : 2 to 1 : 4 depending on the height of the bund and the material used for constructing it. The soil used should be free from roots and twigs which may decay later and cause leakages. The bund is constructed in stages by laying the excavated mud slabs in layers which are compacted and allowed to dry in the sun before adding the layer. Water should be sprinkled over the previous layer before the next layer is laid on, so that there is good adhesion between the layers.

If bunds have to be built in undrainable water logged areas, a hollow frame work in the shape of the proposed bund is erected on the site with bamboo poles, arecanut stems or split coconut trunks and lined with woven palm leaves or bamboo matting to hold the mud which is subsequently dumped into the frame work. By the time the matting decays the mud would have settled and consolidated to form a good bund.

If the soil is sandy and porous, it is mixed with bentonite which has the property of absorbing large volumes of water and swelling upto 8 to 20 times in volume thus plugging the pores in the soil. Chemical sealants, such as, sodium polyphosphate are mixed with poor soil to prevent leakage.

Alternatively to prevent seepage, a clay core is provided in the centre of the bund. A clay core is a wall of clay anchored to the hard substratum below the pond bottom and is covered on either side with sloping more porous soil. In the case of very porous soils, pond liners made of 2-4 mm thick membranes of polyethlene, polyvinyl chloride or buty rubber have been used on the inner surface of the bund.

The outer surface of the main embankment should be provided with granite rubble pitching followed by pointing to prevent erosion. The inner surface of the bunds may be lined with bricks laid on loosely over the soil. The top of the bund and the inner surface are covered with turf to prevent erosion. Slabs of turf are removed along with soil from nearby areas and laid as a mosaic on the surface of the bund. The main embankment could be planted with coconut trees or mangroves.

The bunds in between the stocking ponds need not be as broad as the outer embankment, but should be sufficiently broad to withstand the water pressure. The nursery ponds may by shallow, 30-40 cm deep and small in size 0.05 to 0.10 ha and form about 5-10 per cent of the total water area. The stocking ponds can be 1 to 2 ha in area, rectangular in shape and 0.7 to 1.2 m deep. Smaller ponds get heated up quickly in warm weather and experience wide fluctuations in pH and dissolved oxygen content. Large ponds provide a more stable environment for the prawns.

The nursery ponds can open directly into the stocking ponds through small sluices. Each stocking pond should get its water supply directly from the feeder canal through a sluice. The pond bottom should slope gently towards the sluice gate and could be traversed by shallow radiating or zig zag drainage ditches originating from the sluice gate. These ditches serve to drain the pond at the time of final harvesting and also provide shelter to the prawns during hot days when the surface temperature shoots up.

The bottom of the feeder canals should be lower than the pond bottom so that the ponds can be drained completely, if necessary. The feeder canals should be sufficiently large to supply the water needed by the ponds.

The main sluice on the outer embankment that communicates with the estuary or creek should be sufficiently large and sturdy, preferably made of concrete with manually operated iron shutters. For a small farm a sturdy, wooden open box sluice with two wooden plank shutters sliding in grooves may be sufficient. The stocking ponds can each have a 1.0 to 1.3 m wide concrete or wooden open box sluice fitted with two sliding velon screen shutters to prevent the escape of prawns and entry of unwanted fish and one wooden plank shutter to regulate the water level in the pond.

The land on the windward side of the ponds may be planted with casurina or mangrove trees to act as wind brakes. This will help in reducing erosion of the bunds caused by wind-induced waves.

Economics of Prawn Farming

Prawn farming is a new concept in fisheries. It implies a collective operation of the mechanics and chemistry of the cultivation medium, seed production and sowing, crop management and protection, harvesting and post harvesting activities. With the specific relevance, to prawn farming the above activities can be isolated in four main operations.

1. Farm site development and preparation.
2. Shrimp seed production or procurement.
3. Cultivation operation.
4. Harvesting and marketing.

It is possible to follow the cash-flow through these operations by costing each operation.

Prawn farming is considered as periodic rotation of a wheel with one cycle consisting of one crop. With the development of the farm site, the wheel is ready for rotation. The cycle begins with introduction of prawn seed, the post larvae. The cycle is complete when the shrimps attain marketable size. The time and the distance travelled by the wheel is the cultivation operation. The driving force for the rotation is monetary, that is, the cost of operation. All these components can be costed and watched in financial terms. Thus the farm development as the capital cost and the expenditure incurred through one cycle as the recurring cost may be taken into account. Broad details of the recurring expenditure will be the cost of management, maintenance, seed, feed and the annual depreciation and loan repayment. These will be the inputs per cycle. Outputs will be shrimp production and sale proceeds there of. Debiting the recurring expenditure from the sale proceeds, the net profit is obtained. If the net profit is obtained in this manner is more than 15 per cent of the investment there is no reason why the enterprise of prawn farming should not be attractive.

Cultivable Prawn Species

About 55 species of prawns and shrimps belonging to Penaeidae (27 species), Sergestidae (3 species), Oplophoridae (1 species), Palaemonidae (13 species), Alphidae (4 species), Hippolytidae (1 species), Pandalidae (5 species) and Atydae (1 species) that are either commercially important at present or commercially potential, occur in Indian waters. Several of these species are not suitable for culture. Considering the biology, ecology and distribution, the following eleven species of penaeid prawns are found to be suitable for culture in the salt waters of coastal zone.

1. *Penaeus monodon*

2. *P. indicus*
3. *P. semisulcatus*
4. *P. merguiensis*
5. *Metapenaeus monoceros*
6. *M. affinis*
7. *M. dobsoni*
8. *M. brevicornis*
9. *Parapenaeopsis stylifera*
10. *P. sculptilis*
11. *P. hardwickii*

All the penaeid prawns, though marine in origin, are endowed with a naturally evolved life pattern which is partly spent in the sea and partly in the shallow inshore waters, estuaries and backwaters. The exception to this is *P. stylifera* which completes its life cycle in the marine environment itself. The biological trait of immigrating to coastal waters in the younger stage exhibited by these prawns, their relative faster growth and ability to withstand wide variations of the environmental conditions during this stage of life added to their amenability and advantages of culture in the enclosed waters in the coastal region. The species, like, *P. indicus*, *P. monodon* and *M. monoceros* have wide distributions along Indian coast. But the other species are of local importance as they are restricted to only certain regions of the coast.

But as the intensity of the culture increases, so do the stresses on the life system in the pond, making it necessary to prevent feeding and disease–related problems by doing a thorough job of cleaning the pond and improving the sand quality every season. For these purposes the partial–embankment type is not suitable.

Concerning the intensity of the culture productivity differ in different areas.

Types of Grow-out Pond Prawns / m²	Number of Young Harvested / m²	Quantity of Prawns (kg)
Partial–embankment	20–25	0.4–0.5
Full–embankment	40–60	0.6–1.0
Double–bottomed round tank	70–80	1.5–2.0

**Figure 50: Transportation of *P. monodon*
Post Larvae (wild) for Sale**

**Figure 51: Coastal *P. monodon* Farm
Showing Grow Out Ponds**

Figure 52: Cuttured *P. monodon* Feeding in Check Tray

Figure 53: Harvest of *Penaeus monodon*

Figure 54: Harvesting of *P. monodon* by Pumping and Hand Picking

Low intensity culture utilizing large pond area would lower the running cost. The burden of culture control operations is reduced. By reducing the stress on the population, healthier prawns can be raised.

It is believed that high intensity culture results in not only the increase of excrements, but also an increased interference between individuals in the population that leads to disease and a decrease in feeding efficiency. By strictly maintaining a population intensity of less than 30 PL/m^2 a comparatively high level of production (5.5 tonnes/ha) could be achieved by taking advantage of warm climate through a year–round culture schedule. Seeds are produced twice a year in May and August. By releasing the second crop of post-larva into the same grow-out pond as the first, three months later, the population intensity is kept down to the desired level (Table 14).

Formulated Prawn Feed Used in Farming

Prawns belonging to penaeid group have the habit to eat various kind of food. They used to eat bottom algae and wet feeds like trash fish, minced calms, soyabean cake, peanut cake and rice bran etc. broadcasted by the farmers. But the above foods are either inadequate or tend to decay and pollute the pond water and pond bottom. In farming, therefore, formulated dry feed is used as the main food, while natural food is used as supplementary only.

The highest quality ingredients like fresh fish meal, mixed with balanced plant proteins, wheat, yeast, vitamins and mineral supplements are used to manufacture of formulated prawn feed that are most suitable for the nutritional requirement at each stage of shrimp culture. The food conversion ratio (FCR) has a range of 1.4-1.6 for commercial prawn culture with stocking density varying from 20 to 50 post larvae per square metre.

Having effecting nutrient composition the feed is divided into different categories to suit each stage of shrimp growth (Table 15).

All the categories can be ingested and absorbed by shrimps very quickly resulting in speedy growth. The formulated dry feed has good stability in water. Pellets would last for several hours without changing their shape, which will be totally eaten by prawns. Feed will not contaminate the pond water and pond bottom if given judiciously. As such the shrimps will live in a good environment and the unit of production will be raised.

Table 14: Comparison of the Three Major Methods of Shrimp Culture

	Extensive Culture	Semi-intensive Culture	Intensive Culture
Pond size	5 hectares or larger	1.8 hectares	0.25 hectares
Pond dykes	concrete or earthen	concrete or earthen	concrete
Stocking density (P.L.)	10000–30000 per hectare	60000–1,60,000 per hectare	180000–300000 per hectare
Fry source	Wild	Wild or hatchery	Hatchery
Fry size	PL–20 to PL 35 (25–30 day old post larvae)	PL 20 to PL 35 (20–35 day old post larvae)	PL 20 (20 day-old post larvae)
Water management	Tidal exchange, no aeration	Tidal exchange and pump with aeration water,	Pumps, filter, pre-mixed paddle wheels for aeration and water circulation
Feed used	Natural and occasional supplemental feed	Pelleted supplemental feed and natural feed	Formulated complete feed
Culture period	4–6 months	3–4 months	3–4 months
Harvest size	25 nos per kg	25–30 nos per kg	32 nos per kg
Production (kg/ha/year)	500 or less	3000–8000	10000–20000
Survival rate	50 per cent or less	70–80 per cent	70–80 per cent
Crops per year	1–2	2–3	2.5–3.0

Table 15

Item of Feed/ Content (Per cent)	Baby Shrimp New, No 1 No 1, No 2	Starter No. 3	Grower No. 4	Finisher No. 5
Crude Protein Min	40, 40, 39	38	37	37
Crude Fat Min	3	3	2.8	2.8
Moisture Max	10	10	10	10
Crude Fibre Max	3	3	3	3
Crude Ash Max	16	16	16	16

The moulting of shrimps should take place parallel with the growth to prevent cannibalism. The formulated feed helps the shrimp to recover their strength after moulting and grow uniformly.

Feed are distributed evenly around the pond sides. For ponds bigger than 0.6 ha, it is advisable to feed the shrimps by distributing the feed into inner and outer layer around the pond sides by using raft. The best time of feeding is at 6 AM and 11 PM. The feeding rate is required to be adjusted to the changes of water quality, temperature, weather, remnant feed and prawn appetite.

The quantity of feed to be given is based on the size of the shrimp.

Table 16

Average body weight of shrimp (g)	0.02	0.2	1.5	5	12.5	20	30
Average length (cm)	1	2.5	4.5	7	9.5	12	15
Percentage of body weight	20–15%	15–10%	10–6%	6–4%	4–2%	4–2%	
	Baby Feed No. 1	Baby Feed No. 2	Baby Feed No. 3	Starter	Gro-wer	Finisher	
Baiting time / day	1–2	1–2	2	2–3	4	4–5	
Hours of eating	2.5–3	2.5–3	2	1.5	1.5	1.5	
Water depth (Feet)	1.9	2.2	3.0	3.5–4	4.5	4.5	
Feed in tray net	1/60	1/50	1/45	1/35	1/30	1/30	

The feed is placed in feeding tray at a rate of 8-10 tray per hectare along the dykes and in the feed ration mentioned in the table above.

The trays are inspected daily to check for over feeding or underfeeding. An added advantage of the trays is monitoring of the stock–healthy animals which readily come to the tray. When they look sluggish or do not appear at all on the trays, this means trouble, poor water quality, disease outbreak etc, which needs immediate checking.

Table 17: Feeding Schedule for Prawn Culture

Baby shrimp New No I	6–30 AM (100%)				
Baby shrimp No 1	6–30 AM (50%)			5 PM (50%)	
Baby shrimp No 2	6–30 AM (30%)		3-30 PM (40%)		10-30 PM (30%)
Starter No 3	6–30 AM (20%)	11-30 AM (20%)		5-30 PM (30%)	11-30 PM (30%)
Grower No 4	6–30 AM (20%)	11-30 AM (20%)		5-30 PM (30%)	11-30 PM (30%)
Finisher No 5	6–30 AM (25%)	11-30 AM (10%)	3-30 PM (25%)	5-30 PM (20%)	11-30 PM (20%)

Daily increase in average body length (ABL) and average body weight (ABW) of prawns under different culture days are:

Table 18

Day of Culture	Average Body Length (cm)	Average Body Weight (gm)	Daily Increase in Average Body Length (cm)
P.L stock	1.5	0.05	0.15
10th day	3.0	0.36	0.15
20th day	4.5	1.15	0.15
30th day	6.0	2.65	0.15
40th day	7.5	5.00	0.15
50th day	8.5	7.30	0.10
60th day	9.5	10.00	0.10
70th day	10.3	12.70	0.08

Contd...

Table 18–Contd...

Day of Culture	Average Body Length (cm)	Average Body Weight (gm)	Daily Increase in Average Body Length (cm)
80th day	11.1	17.00	0.08
90th day	11.9	21.36	0.80
100th day	12.7	24.00	0.80
110th day	13.5	27.60	0.80
120th day	14.3	33.5	0.80
130th day	14.9	37.00	0.60
140th day	15.5	42.00	0.60

Identification of quality prawn fry

Healthy fry is the key of successful prawn culture. When choosing the fry, the fry with hairy legs and uropod should be avoided as the sticking *Leucothrix mucor* in legs and uropod may reach gill area and cause asphyxiation leading to death. If the antennules are seperated all the time and cannot even draw close to each other, the fry is unhealthy. In healthy PL 20s abdominal segments should be long, just like the long node of sugar cane, which indicate the fry is healthy and would grow fast. Short abdominal segments is a sign of unhealthy fry. The uropod of PL 20 is always open like a fan while swimming. The wider the uropod opens, the healthier the fry would be. The time from PL 12 to PL 20 is the period that fry put on meat weight. Observing the muscle conditions in abdominal segments the fry can be judged whether it is fully grown up or not. PL 20 fry have the habit of sticking to the sides. It would move quickly to the side and stop moving when removed from pond; otherwise their health would be in a bad condition. By putting a white board with long handle into the bottom of pond, if the fry gather quickly to board, they are well in health.

Changing of shrimp body length and body weight is the direct way of evaluation of the condition of shrimp growth. The corresponding value of body length and body weight are given in the Table 19.

**Table 19: Conversion Table of *Penaeus* Prawn
Body Length and Body Weight**

Body Length (cm)	Body Weight (gm)	Body Length (cm)	Body Weight (gm)	Body Length (cm)	Body Weight (gm)
1.00	0.014	6.00	2.787	11.00	16.625
1.20	0.024	6.20	3.070	11.20	17.531
1.40	0.038	6.40	3.371	11.40	18.470
1.60	0.057	6.60	3.691	11.60	19.440
1.80	0.080	6.80	4.030	11.80	20.445
2.00	0.110	7.00	4.389	12.00	21.483
2.20	0.145	7.20	4.769	12.20	22.555
2.40	0.187	7.40	5.170	12.40	23.662
2.60	0.237	7.60	5.594	12.60	24.804
2.80	0.296	7.80	6.037	12.80	25.983
3.00	0.362	8.00	6.505	13.00	27.197
3.20	0.437	8.20	6.996	13.20	28.449
3.40	0.523	8.40	7.516	13.40	29.787
3.60	0.619	8.60	8.050	13.60	31.064
3.80	0.725	8.80	8.614	13.80	32.480
4.00	0.844	9.00	9.204	14.00	33.334
4.20	0.974	9.20	9.819	14.20	35.278
4.40	1.117	9.40	10.462	14.40	36.762
4.60	1.274	9.60	11.131	14.60	38.287
4.80	1.434	9.80	11.829	14.80	39.853
5.00	1.629	10.00	12.541	15.00	41.461
5.20	1.828	10.20	13.308	15.20	43.111
5.40	2.043	10.40	14.092	15.40	44.804
5.60	2.224	10.60	14.906	15.60	46.540
5.80	2.522	10.80	15.750	15.80	48.320

Use of Theraputants and Chemicals in Prawn Culture Ponds

In aquaculture, the saying "Prevention is better than cure" can be adopted by the farmers to prevent diseases by maintaining the optimal water quality conditions. Through the usual water quality management practices, like water exchange and aerations, diseases can be prevented in extensive and improved extensive farming practices. But these methods alone will not be successful in sustained high production farms using semi-intensive and intensive farming technologies. Application of theraputic agents and chemicals is being practised in order to prevent and cure disease (Table 20).

Health and Disease

In the prawn samples obtained Palk Bay and Gulf of Mannar, some of *Penaeus semisulcatus* were found to be infected with ectoparasite, *Epipenaeon ingens*. This parasite has been earlier described by Nobili in 1906 from Red Sea parasitising *Penaeus semisulcatus*. Dawson (1958) recorded *Epipenaeon elegans* in Persian Gulf from *Penaeus carinatus* (*P. monodon*) and *P. semisulcatus*.

The hosts were found in the length range of 25 to 48 mm in carapace length. There was no correlation between the size of host and the percentage incidence of the parasites.

P. semisulcatus alone was infested by this parasite among many other species of penaeid prawns occurring along with this species. The same condition was observed in the case of *E. elegans* from the Persian Gulf. *E. ingens* was found to occur in the branchial chamber of the hosts, usually in pairs, the male clinging to the ventral surface of the abdomen of the large female. The branchial chamber of these prawns develop a characteristic bulging to accommodate the parasite which completely filled this chamber. The percentage infection of these parasites was more in females (0.23-4.55 per cent) in April, June, September, October and December than male (0.19-2.75 per cent) the maximum being in the last month.

The infected prawns showed degeneration of the primary and secondary sexual organs. In females the ovaries were found always in undeveloped condition irrespective of the size of the host and the season. The petasma of the male also failed to develop to normal size and shape, in proportion to the size of the infected prawn. The parasites filling the entire branchial chamber of the host are likely to produce pressure on the gills surface and reduce the efficiency of respiration. Although the infection, does not cause immediate death, it would probably affect the natural growth of the hosts to a certain extent.

Table 20

Common Name	Commercial Preparation	Purpose of Use	Dosage ppm (parts per million)
Formalin	Aqueous solution of 30-40 per cent Formaldehyde by weight	Control of external parasites including protozoans (Epistylis, Zoothamnium)	15-25
Malachite green	Technical grade zinc free	Control of fungi, external protozoans, bacteria	0.03-0.10
Potassium permanganate	Fine crystals or technical grade	Control of bacteria and external parasites	2-4
Copper sulphate	Crystalline and powder forms	Control of fungi, and external parasites and algal control	Not effective when alkalinity is less than 20 ppm 0.25-0.5 ppm when alkalinity is 20-49 ppm 0.5-1.0 ppm when alkalinity is 50-150 ppm
Chelated copper	Collae and copper control	Algal control	0.2
Simazine	Aquazine	Algal control	0.25
Rotenone	5 per cent emulsifiable concentrate or wettable powder	Fish toxicant	0.05 or 1 of commercial preparation
Filter alum	-	To remove turbidity (decrease pH)	10-20
Gypsum		Decrease pH	250-1000
Hydrated or slaked lime	-	Increase pH remove CO_2	100-1000

The least cared aspect in prawn farming is the health of the animals. The collapse of prawn culture industry in South-east Asian countries was mainly due to the out break of a most dreaded viral disease known as Monodon Baculo Virus (MBV). This disease (MBV) is specific to the tiger prawn, *Penaeus monodon*, which is commercially important.

The diseases of prawn are caused by viruses, bacteria, fungi and protozoans. While some are critical to the life of the animals, others can be cured easily within a few days. The viral disease so far recorded are infectious hypodermal and hematopoietic necrosis (HHN), baculovirus penaei (BP), *Penaeus monodon* type baculovirus (MBV), baculovirus midgut necrosis (BMN), hepatopancreatic parvo like virus (HPV) and reo-like virus (REO). The main sources of all these viral diseases are the sub-clinically infected wild spawners and larvae (seeds).

The fungal diseases caused by the fungi *Lagenidium* spp and *Silopidium* spp are common in the larval stages (hatchery phase) of the prawns, whereas, the fungus *Fusarium* spp attack adult prawns.

The most common bacterial diseases diagnosed in prawns are caused by various species of *Vibrio*. Filamentous bacteria such as, *Leucothrix mucor* and Flexibacter attack the larvae and post larvae of prawns. The protozoans like *Zoothamnium*, *Vorticella*, *Acinata carchesium* and *Euphelota* also infect the prawns.

Treatment and cure is possible with chemicals such as, traflan (for fungi), copper sulphate and cutrine plus (for filamentous bacteria) and formalin (for protozoans) when the disease is diagnosed in time.

However, the viral disease like MBV have no cure. Poor hygienic conditions caused by over intensification (over-stocking of the ponds), poor siting and defective engineering or management practices of the ponds are the major factors for their spread.

At least four virus caused epidemics have adversely affected the global prawn farming since 1980. These viruses are (i) Infectious hypodermal and hematopoietic necrosis virus (IHHNV); (ii) Yellow head virus (YHV), (iii) Taura Syndrome virus (TSV) and (iv) White spot Syndrome virus (WSSV).

Since 1991, shrimp viral disease has reduced production and slowed industrial growth. Many diseases are linked to environmental deterioration and stress associated with intensive shrimp culture. High profits from prawn culture prompted operators for intensive operations that included increased culture densities, increased feed and other inputs, increased waste water loads and increased disease occurrence from various causes.

With the rapid development of penaeid aquaculture industry, many of the significant shrimp pathogens (IHHNV, TSV, and WSSV) moved with live shrimp stocks moved from country to country and from one continent to the other well before their etiology was understood and diagnostic methods were available. Viral epidemics caused major socio-economic losses in shrimp aquaculture from 1986 to 2007.

Since July, 1994, diseases of viral etiology, notably, white-spot disease, have had a disastrous impact on shrimp farming activity of India. Recently, loses due to reported loose shell syndrome have been causing great concern among the farmers. As more than 80 per cent of the farms are owned by small operators, there has been a significant adverse socio-economic impact on small-scale farmers in coastal regions. The latest development in the disease situation in India is the prevalence of Monodon Slow Growth Syndrome (MSGS) as an emerging disease in shrimp culture.

An assessment of loses in shrimp culture due to diseases in India has shown that the epidemic seriousness among the diseases was because of White spot Syndrome Virus (WSSV), Loose shell syndrome (LSS), white gut and slow growth syndrome and combination of WSSV and LSS at national level. The gross national losses estimated in the country are;

1. 48717 tonnes of shrimp production (30 per cent of present shrimp production)
2. Rs. 1022.13 crores in terms of national income.
3. 21.56 lakhs of mandays in terms of farm level employment.

Shrimp aquaculture is not only a foreign exchange earner, but is also a provider of employment (120000 farmers earn their livelihood) and nutritional security of the country.

In India, the major shrimp diseases are, white spot syndrome virus (WSD), Monodon baculo virus (MBV), Hepatopancreatic parvo virus (HPV), Loose shell syndrome (LSS) and Monodon slow growth syndrome (MSGS).

In India, the first report on white spot viral disease in shrimps was from the State of Andhra Pradesh in 1994. With its ability to cause 100 per cent mortality within 5-10 days, the disease caused major economic losses in the shrimp culture industry. WSSV has a wide host range including penaeid shrimps, caridean shrimps, lobsters, several species of crabs, branchiopods, stromatopods, copepods, polychaetes, rotifers and few algae. The virus spreads both horizontally and vertically by live transport of cultured shrimps, from wild species to cultured species, through unhygienic human farm practices, birds and arthropod vectors, infected frozen foods, infected pond water and sediments.

From 1994 to 1999 the shrimp farms on east and west coasts of India experienced repeated out breaks of Monodon Baculo virus (MBV). The virus was detected in *Penaeus monodon* larvae. MBV infection spreads by both vertical and horizontal transmission. Even though the disease has been reported to occur in all the life stages of commercially important prawn species, prevalence and mortality rate were reported to be more in post larva or early juvenile stages than in either zoea or mysis stages. The major signs of MBV are pale bluish grey black colouration of the body, reduced feeding and growth rate, increased epibiotic and epicommensal fouling on the surface of the larvae which leads to sluggish and inactive swimming bahaviour and appearance of white midgut line which can be seen through the abdomen in severely affected larvae and post larvae.

HPV (Hepatopancreatic parvovirus) infected shrimps (*Penaeus monodon, P. merguiensis, P. chinensis* and *P. semisulcatus*) do not show external signs of disease. Stunted growth, poor feeding and size variation in shrimps have been reported with HPV infections. It cause stunted growth and mortalities to the extent of 40-100 per cent in juveniles. Prevalence of HPV in shrimp post larvae have been reported from commercial shrimp hatcheries. Transmission of HPV is believed to be both vertical and horizontal.

The loose shell syndrome (LSS) peaked in the year 2002 in farmed *Penaeus monodon* and the problem continues to persist. Yearly, 40-50 per cent of the farms are being affected by LSS problem with an annual crop loss of 12-25 per cent. The first sign of the disease is the poor feed intake, followed by lethargy. Affected prawns showed reduced growth rate and significant increase in the feed conversion ratio (FCR). Affected shrimp had loose exoskeleton, loose and spongy muscle, stiff rostrum and many shrimp showed discolouration (reddishness). Mortalites, usually occur half way through the grow-out period (40-80 days of farming). The disease has been reported from all farming systems, salinity ranges, pond management methods and varying feed qualities. Higher salinity, high temperatures and algal blooms as the factors coinciding with the occurrence of LSS were cited. The LSS condition could not be cured or controlled using therapeutics.

Monodon Slow Growth Syndrome (MSGS) infected shrimps had stunted growth, poor feeding ability and size variation have been observed in *P. monodon* since last 2 to 3 years in farmed prawns. The most drastic consequences of these features to the farmer is the uncertainty of final harvest output and value. Since no known viral pathogens could be found associated with the MSGS, the causative agent was designated as Monodon Slow Growth Agent (MSGA). The preliminary investigation revealed severe growth retardation and size variation (5 ±15g) after 3 or 4 months farming and the farmers experienced economic loss due to size variation and decrease in yield.

Commercial prawn culture in Asia and Latin America has suffered huge mortalities due to bacterial pathogens belonging to the genus *Vibrio*. The most frequently encountered *Vibrio* species in hatchery and growout facilities are *V. harveyi, V. parahaemolyticus, V. vulnificus,* and *V. alginolyticus.* They are gram negative, motile, rod shaped bacteria that require NaCl for growth. Vibriosis affects all stages of shrimp development but juveniles during the first 30 to 80 days after transfer to the grow-out ponds are especially susceptible. Some *Vibrio* strains are luminescent and infected shrimp often appear luminescent at night. Infections of the exoskeleton often develop into a digestive tract infection, including the hepatopancreas and eventually systemic septicaemia occurs. Many of the hatcheries use antibiotics for the control of Vibriosis without knowing the consequences of the presence of antibiotic residues in shrimp.

Remedial Measures

Besides constant vigil, the disease diagnostics is an essential component in health management of prawn under farming. The earlier methods for disease diagnosis, of pathogens by post mortem, necropsy, histopathology, light microscopy and electron microscopy as well as enhancement and biopsy methods often lack specificity and many pathogens are difficult to detect when present in low numbers or when there are no clinical signs of disease. Direct culture of pathogens from infected tissue and subsequent culture of bacterial strains (purified) and identification is time consuming. The effective control and treatment of diseases of aquatic animals requires access to diagnostic tests that are rapid. DNA based diagnostic methods implies that each species of pathogen carries unique DNA or RNA sequences that differentiate it from other organisms. The techniques offer high sensitivity and specificity and diagnostic kits allow rapid screening for the presence of pathogen. DNA probes for diseases such as WSSV, YHV, IHHNV and TSV are routinely used in many laboratories around the world. DNA probes have also been developed for intracellular parasites and bacteria infecting shrimp.

It has been realized that chemotherapeutic agents are not useful for control of pathogens in hatcheries and shrimp ponds.

The use of probiotics (*Vibrio alginolyticus*) and β–1, 3/6, 6–glucans in *Penaeus vannamei* larviculture was found to influence the survival of shrimp juveniles challenged with WSSV. Certain marine actinomycetes when made available to shrimp post larvae as feed additives have been reported to afford significant protection against WSSV infection. Dietary B-1, 3–glucan (BG) derived from *Schizophyllum commune* is reported to modulate the non-specific immunity of shrimp and increase resistance to WSSV.

A mixture of plant extracts of *Lantana camera, Aegle marmelos, Ocimum sanctum, Mimosa pudica, Cynodon dactylon, Curcuma longa* and *Allium sativum* was shown to be effective in controlling and treating viral infection. The effects of commercially available *Dunaliella* extract (Algro Natural R) on growth and immune functions and disease resistance were determined in *Penaeus monodon*. Shrimp fed on 300 mg of the extract/kg diet exhibited higher resistance to WSSV infection. Total carotenoid and astaxanthin levels were highest in shrimp fed on 200-300 mg of the *Dunaliella* extract.

High Returns

Shrimp farming in Asian countries has enjoyed a boom from mid 1970's. Of all the different species, it is the penaeid prawns, *Penaeus monodon* (Tiger prawn) which had been the prime target of farmers and governments, due to its high market value in the first instance and its export and, consequent foreign exchange potential in the second.

Some shrimp-producing countries have developed their hatchery business to such an extent that they can even meet all the needs of their farms for prawn fry. On the other hand, countries like Bangladesh and India still rely mainly on seed from natural sources. The huge delta of Bangladesh and neighbouring West Bengal, with its numerous rivers and channels, offers an excellent brackish water environment for post-larvae of *P. monodon*. Prawn seed collection has, consequently, developed as a major income generating activity for thousands of people living in these areas of Bangladesh and West Bengal.

Most of the people engaged in prawn post-larva catching in these parts of Bangladesh and West Bengal are landless peasants and poor fishermen, who often have very limited alternatives for subsistence income. Equipped with scoopnets, shooting nets and set bagnets they collect the tiny and fragile post-larvae of *P. monodon* from brackish water sources and sell their catch to middlemen at prices determined by the latter. The fry catcher's desperate need for money makes them an easy target for exploitation.

Before the post-larvae are stocked in the grow-out ponds, they usually go through the hands of one or two more middlemen, are stored for 6-12 hours and counted for a second time before liberation in the grow-out ponds. So, mortality, due to stress and release into an unprotected environment, often soars to 50-70 per cent. The loss is compensated by releasing additional post-larvae and this eventually puts pressure on the natural resource and few existing hatcheries.

Nursery rearing of post-larvae to juvenile size (40-50 mm) greatly increases their viability in the grow-out ponds. The farmer would not require such a large number of post larvae to compensate for loss, and since the grow-out time for juveniles is less, compared to post-larvae, extra crops could be harvested within the season.

Instead of selling the post-larvae right away to middlemen, the nursery would give the fry catchers the opportunity to raise post-larvae upto juvenile size in floating cages, before selling at a price that is reasonable. Juveniles can fetch a price 3-4 times that of post-larvae.

Nursery culturing provides a profitable link between upcoming hatcheries and prawn farms. Hatcheries could sell younger post-larvae to nursery operators, which would improve hatchery efficiency.

Rearing in Floating Cage

The basic design of the floating cage consists of a frame of four narrow bamboo platforms. The sides (measuring 0.45x6.0 m) and the end pieces (0.45x3.0 m) are attached to one another by rope to form a rectangular frame with an inside dimension 2x5 m. Four to six 100 litre plastic barrels are attached under the raft as floats. Two nylon happas (net enclosures) of 1 mm mesh size and measuring 2x1x1 m and one happa of 2 mm mesh size and dimension 3x2x1 m are fixed inside the frame. To prevent folding during tide water movements, the happas are stretched out by frames of galvanized wire and stone sinkers attached to the bottom line. Feeding nets are suspended vertically in the happas and small shrimp and trash fish are minced and applied on them once or twice a day. The entire rearing volume is approximately 8 cubic meter when the upper 20 cm portion of the happas remain above the water line. Stocking capacity is upto 20000 past-larvae, distributed according to size in the three happas.

Net cutting in the happas by mud crabs living in the bottom is one of the constraints met during culture trials. Low tides reduces the distance from happas to the bottom and thus invite the mud crabs to enter the net and cut holes. Sudden and heavy rainfalls changes the brackish water to almost freshwater and this swift change in salinity has been reported as a cause of post-larval mortality. By using happas made from thick threaded nylon net and placing the cages at deep locations the first problem could be solved.

A great variation in survival rates (10-84 per cent) have been observed in cage culture trials. In terms of economic gain, the faster the turnover rate of post larvae, more is the profit. For example, the average gain after 40 days of culture is Rs. 54, for 21 days of culture is Rs. 133 and for 10 days culture is Rs. 120.

The tentative results indicate that nursery culture of tiger prawn post-larvae in cages of the present design and with prevailing market rate is an economically viable activity.

Prawn Farming in India

Table 21: State-wise Details of Prawn Farming (2005–06)

Sl.No.	State	Area Developed (ha)	Area Under Culture (ha)	Production (tonnes)	Productivity (tonne/ha/year)
1.	Andhra Pradesh	79270	69638	53124	0.76
2.	Goa	1001	963	700	0.73
3.	Gujarat	1537	1013	1510	1.49
4.	Karnataka	3435	3085	1830	0.59
5.	Kerala	16323	14029	6461	0.46
6.	Maharashtra	1056	615	981	1.60
7.	Orissa	12880	12116	12390	1.02
8.	Tamil Nadu	5416	3214	6070	1.89
9.	West Bengal	50405	49925	29714	0.60
	Total	171323	154598	112780	0.73

Farming of Marine Prawns in Andhra Pradesh

Prawn farms have been developed in mangrove area of East Godavari district, Andhra Pradesh, adjoining the brackish water creeks, which are rich in prawn seed resources. A very successful 30 ha pump-fed farm has achieved a production of 1000 kg/ha/per crop of 7-8 months duration. Only one crop of *P. monodon* production is feasible in the area as the salinity touches more that 40 ppt during summer months. *P.monodon* seed were stocked at the rate of 50000/ha and harvested at 25-30 gms. size. Ponds were fertilized with cowdung and super phosphate and the prawns are fed with blood clam meat, trash fish, dried prawn head waste etc.

In Amalapuram Taluka, a pump fed 40 ha farm has been developed to culture *P.monodon*, where the annual salinity ranges from 12 to 20 ppt. *P.monodon* seeds are colleted from nearby creeks and stocked in the ponds and fed with rice bran, ground nut oil cake and cotton seed cake. The prawns, on harvest attained a size of 30-40 nos. per kg and the production rate was low at 250 kg/ha/4 months. The low production may be due to the absence of animal protein in the feed given to the prawns.

Paddy field owners of West Godavari district are converting their paddy land with saline soil, not suitable for paddy cultivation in the prawn culture ponds, after obtaining 250 kg of large sized *P. monodon* (20- 25 nos/kg) in 4 months from 0.4 ha paddy land converted in to prawn culture ponds. Subsequently 300 ha of prawn ponds have been developed by the agriculturists with their own resources. These ponds vary in size from 0.5 to 30 ha and have a 10m wide and 60 cm deep peripheral trench and a central platform. The ponds are pump fed, maintaining a depth of 30 to 60 cm over the platform and 90 to 120 cm in the trenches. Cement pipes fixed in the bundhs are used for draining the ponds. The prawn production depends on the salt content of the soil rather than the salinity of the water which is almost fresh except during May-June when it is around 5-10 ppt. As such after the first year the production declined in succeeding years to 200-250 kg/ha/crop mainly due to loss of soil salinity by leaching. In freshwater tank *P. monodon* of 12-15 cm size have been found surviving where the soil is slightly saline.

In Krishna district 45 ha and 72 ha have been developed into prawn ponds on the bank of a creek of mangrove area. Besides private parties have developed another 40 ha in the neighbourhood. Abundant seeds of *P. monodon* (during July-August and October-November) and *P. indicus* were collected from the creeks. In Divi Taluk private farmers constructed 100 ha prawn ponds for culturing *P. monodon*. In these ponds the average production rate was 350 kg/ha/4-5 months in the case of *P. monodon* (20-25 g harvest size) and 450 kg/ha/4 months for *P. indicus* (12-15 g harvest size).

One to three ha size pump-fed prawn ponds have been developed in 200 ha area where the soil is saline and the salinity of water ranges from 0-15 ppt. *P. indicus* culture in a low salinity ponds (4.5 ppt) is the interesting feature of this area. *P. indicus* were stocked as 10 mm post larvae and fed with cotton seed oil cake and rice bran. The harvest of *P. indicus* was 800 kg from 1.6 ha pond in 4 months. The size of the harvest was 50-80 nos per kg.

Modified Extensive Culture of *P. monodon* Using Indigenous Feed

P. monodon post-larvae were stocked at the rate of 5 nos/m² (12mm, 7mg) during 1995 and 3 nos/m² (15mm, 10mg) during 1996. Feeding started on the same day of stocking. A feed with 40 per cent protein was given at the rate of 100 per cent of the body weight for first 15 days. Pelleted feed was provided during the remaining period Ration was reduced to 10 per cent of the body weight during the next one month and thereafter at the rate of 5 per cent and finally to 2 per cent of the body weight.

During the first month of stocking water exchange was done twice a week, and afterwards daily. Shrimps were harvested at the end of 85 days.

Physico-chemical Parameters of Culture Environment

Depth of pond (cm)–70-100

Water temp (0°C)–31.5-33

pH–7.0-8.7

Salinity (ppt)–12-27

DO (ml/l)–4.5-7.0

Zooplankton (Nos/l)–75-120

Zoobenthos (Nos/m²)–100-123

Feed ingredients (per cent)

Soya powder–35%

Prawn shell powder–10%

Fish meal–10%

Gingelly oil cake–3%

Coconut oil cake–10%

Vegetable oil–5%

Vitamine mixture–1%

Mineral mixture–1%

Rice bran–5%

Tapioca starch–20%

Culture Details

Area of pond (m²)–400

Rearing period–85 days

Stocking density–5 and 3 nos/m²

Initial size–(mm/mg)–12(7) and 15 (10)

Initial biomass–(g)–14 and 12

Final size (mm/mg)–150 (20) and 160 (26)

Survival rate–73.25 and 83

Qty of feed give (kg)–57 and 47

Qty of shrimp harvested (kg)–29 and 26

Food conversion ratio–1.86 and 1.61

Gross production rate (kg/ha/85 days)–732 and 647

Expenditure (Rs)–1875 and 1448

From sale of prawn–3516 and 3237

@ Rs. 120 and 125/kg

Net profit–(Rs.)–1641 and 1792

Organic Prawn Farming

Marine Products Export Development Authority (MPEDA) jointly with the Swiss Import Promotion Programme (SIPPO) adopting organic prawn farming standards from Naturland, Germany and Indocert, the local certification agency, started organic farming of paddy and prawn in Kuttanad in Kerala. The results were encouraging. Although the paddy production was found to decrease about 20 per cent under organic farming, it was compensated by more than 33 per cent increase in gross income and 36 per cent higher net profit. The organic prawn yield also fetches an increased price by 20 per cent compared to conventional prawn yields.

In Kolleru ponds of Andhra Pradesh, Organic scampi (*Macrobrachium* spp) production in 2008 was of the order of 480 kg/ha as certified by Naturland, Germany.

Great progress has been achieved in the usual giant freshwater prawn (Scampi) farming all over the world, especially since 2000. The present production of farm gate values exceed 400000 tonnes/year and US $ 1.75 billion respectively. China has been leading in the scampi production for quite some time with over 300000 tonnes/year, while India, the second largest producer in the world produces only around 30000 tonnes.

In Kerala, successful harvest of three organic paddy–prawn crops has been encouraging. Average yield of about 400 kg/ha and a total of 3.5 tonnes of scampi was obtained from three farms.

In Andhra Pradesh organic farming trials was taken up by two farmer societies comprising about 25 farms covering an area of 30 ha, a higher yield of upto 800 kg/ha is expected.

In the first phase of the Indian Organic Aquaculture Project (IOAP) only the two states, Kerala and Andhra Pradesh have been included. Where organic paddy and prawn farming are linked, it is the paddy crop which is certified first and accorded the status of a conversion period, when the organic farming standards are first applied to the field. This phase marks a gradual transformation of the field which has been using toxic chemicals, pesticides, and inorganic fertilizers over the years to an organic realm which abstains from all such chemicals and when the farmer trains himself in the use of organic inputs based on a system of integrated pest and nutrient management to harvest the first crop of paddy in the transitional phase.

In the second year, the farmer who becomes more confident applies the organic farming standards fully in his farm and produces the first crop of organic rice which may be certified by the Agency, once fully satisfied by the conditions prevailing. Organic prawn production will also be in the transitional stages during these phases and only in the third year that prawn farming will be certified as organic if the farm fully satisfies the guidelines of IOAP. Such adherence to the standards is ensured by frequent visits of technical staff from Indocert, and based on proper farming data being maintained by the farmer. Maintenance of proper farming records in the farmer's diary is a mandatory requirement in the Project based on the Internal Control System (ICS) developed, by which the day to day activities in the farm are diligently recorded by the farmer and closely monitored by a panel of Field Inspectors. This effectively ensures that organic farming principles are properly followed throughout the farming period.

Marketing of the organic prawn is being made based on a contract with the certified processor who would have agreed to purchase the produce at a 20 per cent premium above the prevailing market rate. This has the twin advantages for the farmer who gets a better price and avoiding any middlemen in marketing, since the direct and timely payment by the processor is assured.

Naturland, Germany, in consultation with Indian experts have established a set of guidelines for proper implementation of the Organic Aquaculture. Project, which covers almost all the relevant phases in production of both paddy and prawn. The Agency ensures by way of farm inspections at different stages of production that these guidelines are properly followed.

General Guidelines for Food Safety; Hygiene and Social Responsibility

1. Wild life is protected and not harmed / killed. Only netting and scaring against predatory animals is done.
2. At least 50 per cent of the dyke surface is to be covered by vegetation.
3. Coconut trees, banana and other plants on dykes are to be managed organically.
4. No trees to be cut in the farming area.
5. Farm surroundings to be kept clean.
6. No scrap plastic materials, such as, bottles, bags, nets etc to be allowed in the pond/farm premises.
7. Ducks should not be grown during prawn culture in the same pond.
8. Entry of cattle, dogs etc into the pond during grow-out should be prevented.
9. Wages for employees should be based on the minimum wage-guidelines approved by the concerned State Government.
10. Child labour is not permitted.

Paddy and Prawn Production

1. Organic prawn culture to be done after harvesting of paddy crop.
2. Paddy must be certified as organic.
3. No agro-chemicals (like herbicides, fungicides and insecticides) and inorganic fertilizers to be used during paddy cultivation.
4. Only rotational crop of organic rice and prawn is allowed.
5. Maximum stocking density for organic prawn farming is fixed at 25000 PL/ha.
6. Additional stocking of seed of Indian Major Carps and grass carp could by @150 fingerlings/ha.
7. Organic prawn seed to be procured from certified hatcheries only.
8. Feed taken only from certified feed mill to be used.

9. Only certified organic products such as probiotics should be used and

10. No antibiotics and other chemicals to be used in pond and water management.

11. Water from the adjacent river/canal could be the source for prawn culture.

12. Water quality parameters to be monitored periodically.

13. Water can be let into the pond by gravity or by using a pump, filtered by net.

14. The pond has to be sun-dried.

15. Only mahua oil cake, tea seed cake and *Croton tiglium* should be used for eradication of undesirable species.

16. Only agriculture lime or natural dolomite permitted to be used for maintaining pH.

17. Only compost from organic rice production, sun-dried cattle dung should be used for fertilization.

18. No fertilization is to be done post-stocking, except in emergency (algae crash) and

19. No urea, NPK/other inorganic fertilizers or chemicals/disinfectants to be used in the pond.

Postharvest Handling and Marketing

1. Prawns to be washed and iced immediately after harvest.

2. Only quality ice from a certified processing plant to be used.

3. Only boxes from certified processor should be used.

4. No beheading and removal of claws should be done at the farm site and

5. Marketing is the responsibility of the certified processor.

Chapter 12

Adoption of Improved Technology in Prawn Farming

Collection of Parent Prawns

Loss of farmed prawns due to viral diseases like white spot disease, loose shell syndrome etc is a major concern for the industry. Use of specific pathogen free seeds for stocking is viewed as a strategy to avoid diseases. It is believed that use of disease-free spawners for the purpose of breeding to raise seed would greatly reduce the chances of incidence of viral diseases in farms. Matured prawns collected from the coast of India are known to be carriers of the viral pathogens; while the brood prawn collected from waters around Andaman is generally believed to be disease free, and thus offer a good scope for the local inhabitants to earn livelyhoods from the industry.

There are about 60 families of fishermen living adjacent to the coastal belt of Betapur, located on the eastern coast of the Andaman Sea in North Andaman District, which is about 190 km from Port Blair and about 22 km from Rangat, a township in north Andaman. The fishing ground lies alongside the pollution free Culbert Bay. The fishing area of tiger prawn brooders is restricted to about 2-3 km from north to south and about half km towards the sea at 8-10 m depth region. In the northern edge, a local stream, Dhani Nala, empties copious quantity of freshwater into the sea, diluting the seawater in that area. The area is quite shallow with a muddy bottom. The spawners of tiger prawn migrate to this area for spawning because of congenial environmental conditions and bottom topography.

Considering the available potentialities in the region, Andaman and Nicobar Administration established a shrimp brood stock bank and nauplii production centre for which a facility was developed at the seashore site, where live *P. monodon* spawners are induced to spawn and conducive conditions were provided for the fertilized eggs to hatch into nauplii for transportation to the main land hatcheries for raising them into post larval stages.

The fishermen living adjacent to Betapur, during the brooder collection season; from June to November, go for catching tiger prawn spawners. The peak season of brooder collection is August to September. The spawners are said to be available in December also. But due to the roughness of the sea during the month, the fishermen are unable to go out for fishing in their small boats. Of thirty five to forty traditional boats operating in the region, about 25 boats go to the sea in the early morning at about 3 Am for catching spawners. There are atleast two fishermen on each boat. After about 30 minutes of searching, they lay outs the nets in the specific area. Trammel net, locally known as disco net, consisting of two nets of different mesh sizes, joined together, one over the other to form an entangling net are used to catch the spawners. The length of the net is 120-150 feet and its depth is 9-10 feet. The total cost of the complete gear with floats, sinkers, head and foot rope is about Rs. 3000/-. Generally the fishermen carry more than 10 nets per trip. Many fishermen do not have their own nets and they hire nets from others. In that case the harvest is shared between the fishermen and the net owner at the ratio of 1 : 2.

Shrimps thus caught in the net are placed in the boat with sufficient quantity of water. No aeration or any other precaution is taken on board. After arrival in the shore, shrimps are immediately transferred to an exclusive place, where the spawners are placed in plastic tubs in a very low stocking densities with aeration, till the required number of brooders are collected. The brood shrimps are transported to Port Blair in rectangular (30x45x80 cm) polyethylene brooder bags with rostrums covered by rubber tubes and sharp edges of walking legs slightly trimmed off. Each bag with 10-12 litres of water can carry about 6-9 spawners depending on sex, size and their maturity stages. They can be transported in such bags for 8 hours to reach Port Blair without any mortality.

Generally the spawners are selected by the hatcheries based on the weight, stages of maturity and moulting stage of the prawns. Hard shelled female tiger prawns at advanced stage of maturity, weighing 120 gram (more than 100 gram) and males weighing 70 to 80 gram each are generally preferred, while the smaller and immature shrimp are rejected. The male and female live spawners fetch a rate of Rs. 150 and Rs. 350 per piece respectively, which however change based on the availability and demand of brooders. Both in the disco netting, as well as in regular fishing activities, the fishermen get good catch of small sized *P. monodon*, *P. marguiensis*, *P. semisulcatus* and other small shrimps along with fishes. They are sold in the local market in fresh condition. *P. monodon* fetches Rs. 200-250 per kg., while other shrimps fetch Rs. 120-180 per kg.

Andaman and Nicobar Administration allowed two private mainland hatchery operators to set up nauplii production centres at Andaman, in order to conserve wild tiger prawn brooders. The Administration has also set up a maximum limit of collection of brooders from the wild.

Improved Hatchery Technology

In recent years the development of shrimp farming from a low level traditional subsistence activity into a commercial enterprise has seen tremendous transformation in the brackishwater sector. The pioneering works of Motosaku Fujinaga (Hudinaga) in 1935 with the successful spawning of kuruma shrimp (*Marsupenaeus japonicus*) in controlled conditions led to the growth of shrimp farming worldwide. At present, about 1,43,000 ha of brackish water is under shrimp farming in India, with an annual production of around one tonne per hectare. The two major inputs for successful shrimp farming are seed and feed. Shrimp aquaculture in India is essentially one of farming of the black tiger shrimp, *Penaeus monodon*. The demand for seed of this species is felt very high and nearly 30 billion post larvae will be required to meet the seed requirement by 2010.

The establishment of two commercial shrimp hatcheries with foreign technological collaboration by Marine Products Export

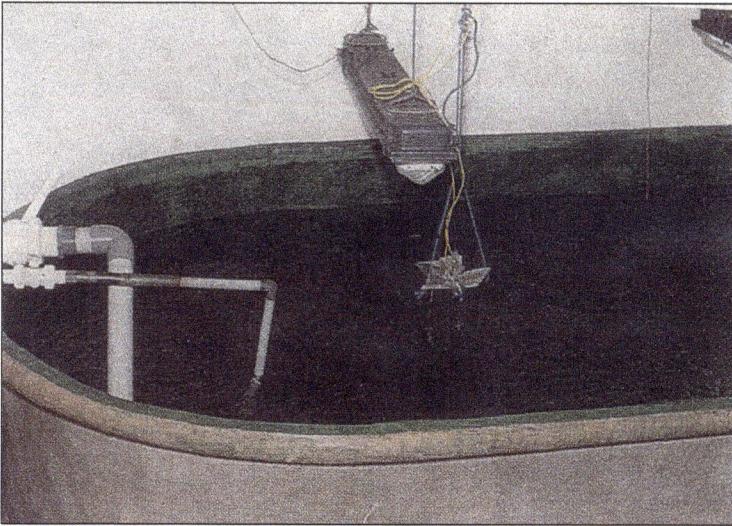

Figure 55: Larval Rearing Tank

Figure 56: Masonary Larval Rearing Tank

Development Authority (MPEDA) in 1980s was the turning point in the history of Indian shrimp aquaculture. This was a milestone that

Figure 57: Prawn Seed

Figure 58: Algal Culture

Figure 59: Polythene Lined Earthed Rearing Pond

Figure 60: Artemia Hatching Tank

opened up the flood gates of shrimp aquaculture development in the early 1990s in the country. Thereafter, with the entry of several potential stakeholders, starting from small scale and marginal farmers to entrepreneurs and those from corporate sector, coastal and brackish water aquafarming sector witnessed exponential growth. This was made possible by the establishment of state of art hatcheries by the private sector along both the coasts of India, inspired by the initiative of MPEDA in the setting up of the two model commercial shrimp hatcheries in the country, now with combined production capacity of 12 billion seed.

Nature of Hatcheries

Based on the production capacity, hatcheries are categorised as below;

1. *Backyard hatchery* are hatcheries with an annual production capacity of 10 million seed. Most of this type of hatcheries came up in Kerala for production of *Penaeus indicus* seed.

2. *Small scale hatcheries* with 20 million seed production capacity per year come under this category. These hatcheries produce *Penaeus monodon* seed.

3. *Medium scale hatcheries* produce 50 million *P. monodon* seed annually;

4. *While large scale hatchery* produce more than 50 million *P. monodon* seed annually.

A total of 351 shrimp hatcheries are now distributed throughout India with production capacity of 14302 million seeds as detailed below; Gujrat–2 (45 million); Maharashtra–8 (345 million); Karnataka–14 (321 million); Kerala–29 (537 million); Tamil Nadu–81 (3078 million); Andhra Pradesh–191 (9335 million); Orissa–15 (475 million); and West Bengal–11 (166 million).

Suitable Sites

Selection of suitable sites is important for viability and sustainability of shrimp hatchery. It should be located close to the quality seawater source with continuous supply of clean and clear sea water. An ideal site is one located adjacent to the coast bordering calm sea and away from any freshwater or brackishwater inflow. Sea water temperature between 25° to 32°C without wide fluctuations is optimal for most of the tropical species of shrimp. The salinity should be between 30 and 34 ppt. Cyclone prone areas and areas of continuous heavy rainfall or freshwater influx from rivers should be avoided.

The other parameters that affect the growth and survival of the shrimp larvae together with their optimal levels are; temperature– 28-32°C (18°-36° tolerable limits); salinity–30-34 ppt (26-34 ppt tolerable limits); pH–8.0-8.4 (7.0–9.0 tolerable limit); dissolved oxygen–Above 4 ppm (Above 3 ppm tolerable limit), ammonia–N– less than 0.01 (upto 0.1 ppm tolerable limit) and nitrite–N–less than 0.01 ppm (upto 0.1 ppm tolerable limit).

Hatcheries generally depend on the wild spawners. As such it is essential to locate the hatchery near the landing centre. Availability of infrastructure facilities are also to be considered for approach roads, electricity etc.

Design and Construction

A well-designed shrimp hatchery would have to consist of independent facilities for quarantine, brood stock maintenance, induced maturation, spawning, hatching, larval rearing, indoor and outdoor algal culture and for *Artemia* hatching.

Hatcheries have to draw sea water from borewells or directly from open sea by constructing concrete jetties into the sea beyond the wave breaking zone. Non-corrosive materials, such as, plastics, PVC, concrete and wood are to be used for sea water supply.

Sea water has to be filtered through sub-sand well points, sand filters (gravity or pressure) and or mesh bag filters into the first reservoir or settling tank. Following settlement, the water is disinfected by chlorination followed by dechlorination by sodium thiosulphate. Thereafter it is filtered through cartridge/bag filter and finally disinfected using ultraviolet light and/or ozone. The use of activated carbon filters, addition of EDTA and temperature and salinity regulation is to be followed to ensure supply of quality water into the hatchery. Water discharged from the hatchery (contaminated) should be held temporarily and treated with hypochlorite solution (more than 20 ppm active chlorine for more than 60 minutes).

Pumps made of cast iron and stainless steel are ideal for saline environment. The capacity of the pump is calculated based on the total tank capacity, maximum water requirement per day and time limit for water exchange in tanks.

PVC pipes and pipelines are used for the water supply to the hatchery. There should be independent inlets for each tank with valve arrangement.

Air blowers and compressors are used to supply air to all tanks to increase dissolved oxygen level, for stabilizing dissolved organic matter and to provide sufficient turbulence to maintain uniform suspension of both the larvae and the feed materials. Air supply is done through PVC and polyethylene pipes at a uniform level in all tanks. In the larval rearing tanks air delivery would have to be either through diffuser stones or through air-lift pumps.

The tanks in the hatchery (reservoirs, overhead tank, maturation tanks, spawning tanks, larval rearing tanks, nursery tanks or post larval rearing tanks, and algal culture tanks) are generally rectangular, circular or oblong in shape or may be plastic lined with aluminium frame, fibreglass reinforced plastics, concrete hollow blocks and reinforced concrete. Concrete tanks are coated with epoxy paint to provide smooth interior surface. All angled parts of the tank should be rounded off to facilitate cleaning.

Reservoirs provide water need of the hatchery at the time of greater water demand over a relatively short period beyond the capacity of the pump. It is used for the chemical treatment of seawater. Overhead tank provides continuous water supply to different tanks by gravity.

With a water capacity of 5-10 cubic metre and 1 metre effective depth, maturation tanks are housed in closed sheds under darkness to avoid disturbances to the shrimps from human movements. Spawning tanks are of 250-500 litre capacity. Larval and post-larval rearing tanks are of 10-50 m^3 capacity with an effective water depth of 1-1.5 m, housed in a shed with reduced light conditions to prevent heavy growth of algae. Depending on the scale of operation and daily requirements, algal culture tanks may be of 1-10 m^3 with effective water depth not more than 1 m to allow light penetration through the water column.

Hatchery building need not be totally enclosed. The larval rearing tanks is to be shielded from direct sunlight and rain. The building should provide space for the living quarters of the technicians, for a phycology laboratory, feed preparation room, a laboratory for water quality and biological analysis and a packing

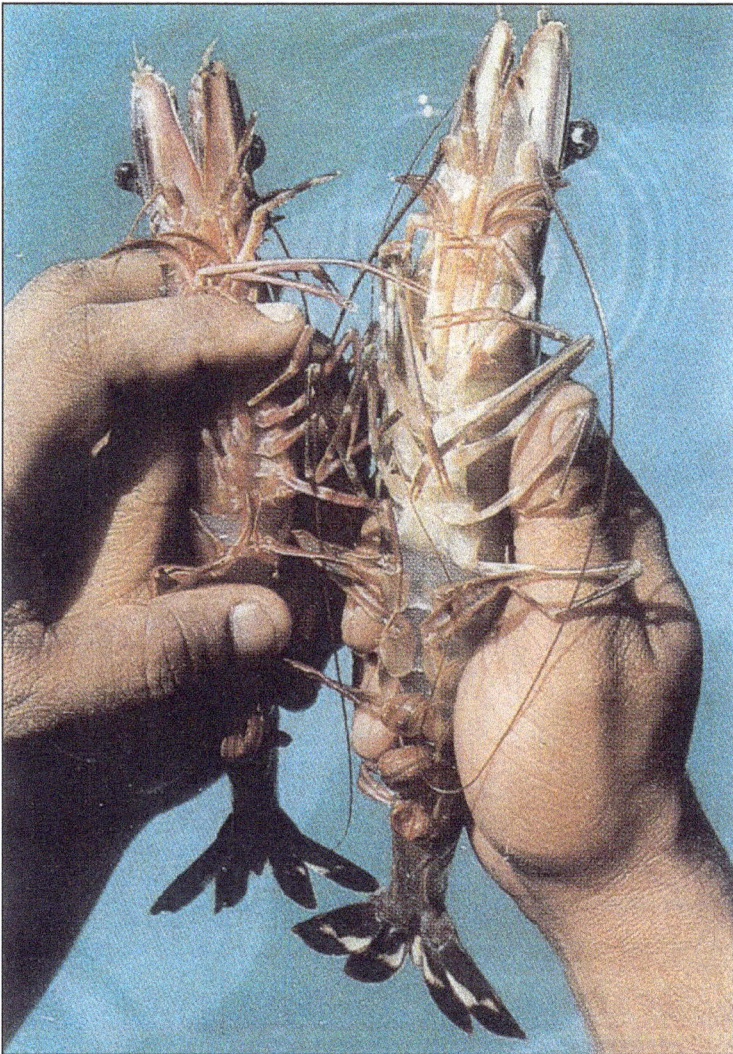

Figure 61: Differentiating Male and Female *P. monodon*

room. Pumps and filtration units in the seaward side and blowers and generators on the landward side are located in the seperate buildings. There should be space for free movement, optimal use of water and air supply system.

Figure 62: Broodstock Rearing Tank

Figure 63: Maturation Tank

Spawners

Mother shrimps are transported in sterilized sea water into the hatchery to prevent contamination.

Availability of *P. monodon* spawners from sea is seasonal, limited and often erratic. To overcome this problem *P. monodon* brood stock is raised in earthen ponds by rearing post larvae 20 (PL 20) at the rate of $1/m^2$ for 8-10 months using natural feed and pellet feeds at the rate of 5-10 per cent of their total body weight. Water management includes 10-30 per cent exchange of sea water every day. Alternatively PL 20 are reared upto 30-40 gram weight each at a density of $3/m^2$ for 5 months and rearing them up to 80-100 gram at density of $1/m^2$ for 5 months using fresh feed at the rate of 10 per cent of their body weight and supplemented with pellet feeds at the rate of 5 per cent of the total body weight and with 10-50 per cent of water exchange.

Healthy, disease-free female *P. monodon* above 90 g and males above 60g can be induced to maturity. However, larger males above 80 g will ensure higher fertilization rates. Similarly females above 100 g will maximise egg production per spawner.

For transportation, the temperature in the broodstock holding tank is gradually reduced to 18°-28°C at the rate of 1°C for every 10 minutes and transported in double polyethylene plastic bags placed in thermocole boxes. A rubber tube is inserted over the rostrum to prevent puncturing of the container bags during transportation. Dissolved oxygen level more than 5 ppm are maintained by filling one third of each bag with chilled filtered seawater and two-third with pure oxygen, bubbled into the water.

Arriving at the hatchery, the spawners are held in the quarantine holding area and slowly acclimated to the hatchery water condition. They are given 100 ppm $KMnO_4$ bath for 30-60 seconds and then released at the rate of 2-3 prawns per square metre into the maturation tank. They are provided with fresh, high quality feed that includes polychaete worms, squid and bivalve molluscs or enriched adult *Artemia* and dry broodstock diets at 25-26 per cent of body weight divided into 6-8 doses per day. The temperature in the receiving tank is gradually increased at 1°C per hour to match the ambient temperature, while salinity is adjusted at the rate of 1 ppt every 10 minutes. Water exchange is done at 200-300 per cent per day. The optimal water quality parameters required for maturation are 28-29°C, salinity of 30-35 ppt, pH 7.5-8.5 and NH_3 ammonia and NO_2 nitrate at less than 0.1 ppm.

Induced Spawning

Hard shelled, inter moult healthy females free from disease or injury having spermatophore in the thelycum are selected for eyestalk ablation. Unilateral eyestock ablation is done either by cutting, incision and pinching or by electrocauterisation. The use of electric battery is the best way of ablating the eyestalk since it causes minimum stress. Ablation should be done as quickly as possible to minimise the stress. After ablation, the area around the cut eye is disinfected with pure liquid povidone iodine solution and the animals are released into the maturation tanks along with unablated males at the rate of 4 nos per sq. metre in the ratio of 2 females : 1 males to ensure best mating success. Mating generally take place in the night. The ablated shrimps are fed with fresh feeds such as calm, squid, mussel meat in rotation at the rate of 15 per cent of the total body weight distributed four times a day, while polychaete worms (6 per cent of the biomass) or *Artemia* (3 per cent of the biomass) are given once in a day. Pelleted feeds with 50 per cent protein and 10 per cent PUFA can also be given with fresh feeds. The optimal environmental conditions required for maturation are,

(i) Housing–ventilated roofed shed; (ii) Tank size–5-15 tonne capacity, circular or rectangular made of fibre glass or concrete, (iii) Light intensity–Reduced 100 lux (artificial) dim light; (iv) Light quality–Blue or green; (v) Photoperiod–12 hour light, 12 hour dark; (vi) Water depth–80-100 cm; (vii) Salinity–30-36 ppt; (viii) pH–8.0-8.5; (ix) Dissolved oxygen–Saturation by continuous aeration; (x) Stocking rate–4 nos/m^2; (xi) Stocking size-Females–90-180 g and males 60-90 g; (xii) Sex ratio–2 females : 1 male; (xiii) Water management–100 per cent exchange/day using filters, 200 per cent exchange by flow through system/day; (xiv) Fresh feed–Calm, mussel, squid, oyster meat at the rate of 15 per cent of the total biomass/day, polychaete worms at the rate of 6 per cent of the biomass or *Artemia* biomass at the rate of 3 per cent of the total biomass/day; (xv) Artificial pellet feed 2 per cent of the total biomass/day; (xvi) Feeding schedule–Four times a day.

Females with IV stage ovaries are spawned individually in 500 litre spawning tanks. After spawning the females are returned to the maturation tank. The spawned eggs are collected by siphoning, cleaned with seawater, disinfected and transferred to 10 litres basins. After thorough mixing, two 10 ml samples are taken and the number of eggs present is counted and computed to the total volume of water, to get the total number of eggs released per spawning. About 100 eggs are examined under a microscope to determine the quality of eggs. Wild females have fecundity of 2,00,000 to 10,00,000 eggs depending on the size of prawn with an average of 4,00,000 eggs/females. But ablated females of 90-150 g weight spawn 1,00,000 to 5,00,000 eggs each, with an average of 2,00,000 eggs/shrimp. The fecundity and quality of eggs decrease with repeated spawning.

Washing and Disinfection of Eggs

Eggs are collected gently in a 50-60 micron mesh net and rinsed gently in running sea water for 5 minutes. The eggs are then dipped in 100 ppm formalin solution for 30 seconds and for one minute in 50 ppm povidone iodine solution and thereafter gently rinsed in running sea water for 5 minutes. After that the eggs are stocked in hatching tanks.

The eggs from each spawning tank are transferred to seperate hatching tanks filled with clean sea water with mild aeration. 5 to 30 ppm EDTA and 0.05-0.1 ppm treflan are also added to the water. The eggs hatch out into nauplii in 12-16 hours after spawning, depending on the water temperature. The nauplii from each spawning are estimated by counting four 100 ml random samples and multiplying the numbers to the total volume of water. The hatched nauplii that come to light are harvested and disinfected before stocking into the larval rearing tanks.

Washing and Disinfection of Nauplii

Nauplii are gently collected in a 100 micron mesh net and rinsed in running sea water for 5 minutes. They are then dipped in 200 ppm formalin solution for 30 seconds, followed by one minute dip in 50 ppm povidone iodine solution and thereafter rinsed in running sea water for 5 minutes. Nauplii are then stocked in larval rearing tanks at a density of 100-150/litre filled to 50-75 per cent capacity with clean disinfected filtered sea water of 28°-30°C and at 30-35 ppt. During the zoeal stages water is not changed, but water added daily until the tank is full at early *mysis* stage. Water exchange is done daily at 10-30 per cent during *mysis* stage, 30-50 per cent during PL 1 to 5 stage and more than 50 per cent from PL 6 stage until harvest at PL 15 stage. If disease or water quality problem occurs, water exchange rate should be increased. The water quality parameters, namely, temperature (28°-30°C), salinity (30-35 ppt), pH (7.8-8.2), NH_3–ammonia and NO_2 nitrate (less than 0.1 ppm) should be monitored daily in each tank. Uneaten food and faeces are siphoned out from the bottom of the tank periodically.

It takes around 8-12 days from nauplii to PL I and from PLI to PL 20 another 20 days and the whole larval cycle is completed in 30-32 days. Generally 6-8 larval cycles can be taken in a well managed and maintained hatchery.

Feeding

Live *Chaetoceros, Thalassiosira* or *Skeletonema* sp maintained at 80-130000 cells per ml or preserved algae are fed to zoea and *mysis,* frozen/dead *Artemia* nauplii only for *mysis* stages and live *Artemia* nauplii for PL stages. Long faecal tails and full digestive tracts indicate that the zoea is healthy and feeding. First instar *Artemia* nauplii disinfected and killed by freezing are fed to *mysis* stage larvae and supplemented with dry or liquid diets of 50-150 micron sized particles, while live first instar *Artemia* nauplii is fed to postlarval stage shrimp. Dry, liquid and crumbled flake diets of 200-300 micron sized particles for PL 1-8 and 300-500 micron from PL 9-15 should be fed 6 times daily alternating with 6 feedings of *Artemia* per day. Microparticulate, microencapsulated and microbound diets are commonly used in hatcheries. They are 5-2000 micron in sizes and can be used as diets for different larval stages.

Collection and Transportation

Harvest and transportation of PL should be done gradually and with minimum stress to ensure good survival rate on stocking into the grow-out ponds. Salinity adjustment is initiated from PL 10 by adding freshwater to achieve a salinity change of less than 3 ppt per hour from 30-20 ppt, less than 1 ppt per hour from 20-10 ppt and 0.5 ppt per hour from 10-5 ppt. After harvesting, the PL are held at the rate of 1 million per cubic metre in tanks containing clean disinfected and filtered seawater with continuous aeration. Temperature is gradually reduced from 28°-30°C to 23° C over 30-40 minutes to bring down stress. Seed are transported in plastic bags at the rate of 500 to 1200 PL per litre and usually held inside thermocole lined cardboard boxes. Live *Artemia* nauplii are added at 15-20 /PL for each 4 hours of transportation to provide food to larvae.

Algal Culture

Live diatoms and microalgae (*Chaetoceros, Thalassiosira* or *Skeletonema* sp) are preferred shrimp larval diets since they offer perfect nutrition, self-suspending in the water coloumn enhance water quality by absorbing ammonia, nitrite and carbon dioxide etc, maintain shed in the water and produce natural and helpful antibiotics. So mass culture of these microalgae is a pre-requisite for successful hatchery operation.

For indoor as well as outdoor mass culture of algae, filtered seawater free of toxic or unwanted sediments is a pre-requisite. Sterilized glassware and other materials are used during pure culture and the culture flasks or jars are covered with aluminium foil or cotton to minimize contamination through air. The culture can be differentiated as follows: (i) In maintenance culture, natural collection kept in culture containers; succession of dominant species takes place. (ii) In enrichment culture, crude collection, treated with selected media that may favour the rapid increase in number of desired species. (iii) Population of a single algal species in unialgal culture is done with associated microorganisms. (iv) In axenic culture population of a single algal species is maintained without any other living organisms. In clonal culture population of organisms descend asexually from single individual.

Algal culture should be maintained in pure form in the indoor, temperature controlled rooms and be used as the starter source for outdoor mass culture. The population growth of algae is characterized by a sigmoid curve and is divided into 4 distinct phases, the lag phase, logarithmic or exponential phase, stationary phase and death phase. Algae in the exponential growth phase are fed to the larvae. From the outdoor mass culture the algal water is pumped into the larval rearing tank or filtered through a fine mesh bag and the concentrate is then added to the larval rearing tanks. In both cases, the cell densities (for *Chaetoceros* sp.) are maintained at 80-130000 cells per ml for zoea and mysis (peaking at z_3) and 50-60000 cell/ml during early PL stages.

Artemia is the most widely used live food for larviculture in aquahatcheries world wide since of its ready availability in the form of dry cysts containing dormant embryo. The dormant cysts can be hydrated and hatched into nutritious nauplii within 24 hours enabling supply of live food "on demand". Though *Artemia* cysts can be hatched directly into nauplii, it is desirable to decapsulate the eggs to improve the hatching percentage and also to eliminate the chances of disease contamination.

Health Monitoring

Each larval tank is monitored twice daily to assess health of the larvae and to take corrective measures for any problem noticed. This is crucial for takings decisions on water exchange, feeding and other management activities. Larval samples are examined for their stage, swimming activity, behaviour, feed intake, muscle–gut ratio, excretion, disease or physical deformity. Samples from broodstock, eggs, nauplii, post larva should also be screened for viral diseases like white spot virus. Monodon baculovirus. Periodic examination of water and larvae are also done to find out luminescence bacterial disease, necrosis etc. Presently techniques are being evolved to produce seed by using probiotics in the larval rearing phase, in various stages of the larval cycle mainly through water.

To secure a disease free environment in all production phases and to prevent losses from disease infection through effective elimination of pathogens and theirs carriers, which are believed to be transferred between regions through the importation of broodstock, post-larvae and shrimp products, the measures and methods of biosecurity have been adopted, which in a shrimp hatchery includes;

1. Specific pathogen free (SPF) and Specific pathogen resistant (SPR) shrimp stocks should be used.
2. All incoming stocks should be analysed for disease and quarantined in the designated area.
3. All incoming water sources should be treated to eliminate pathogens.
4. Equipment and materials should be sterilized and maintained clean.
5. Personal hygiene measures including washing of hands and feet and clothing should be enforced.
6. Knowledge of the potential pathogenic diseases and the source of risk and methods and techniques for their control and for eradication should be ensured.
7. Maintenance of optimum environmental conditions has to be ensured.

    8. Immune enhancers and probiotics are to be used in the place of antibiotics.

Responsible Prawn Farming

The long-term success of shrimp farming depends upon providing a good growth environment in culture ponds for shrimps. Environmental conditions in shrimp production ponds are directly linked to the ecology of the coastal zone. Therefore, it is in the best interest of shrimp farmers to use environmentally responsible production practices that will not have any negative impacts on the environment. In recent years, increasing attention has been given towards sustainable management of shrimp farming through discussions and interaction between experts and farmers, particularly in the light of the serious shrimp viral disease problems, strict food-safety standards of the importing countries and increasing environmental concerns.

Management strategies for sustainable shrimp farming can be done at two levels, one is at the level of the farm which involves proper utilisation of resources and inputs. These practices have to be followed by the farmers. The other is at the level of policy makers so as to integrate shrimp farming in the overall development plan of the coastal zone. The farm level strategies are;

(a) Most of the ill effects of shrimp farming can be attributed to poor site selection (in close proximity to each other) in India. The major issues of environmental impacts of shrimp farming, among others, consist of (i) destruction of mangroves, (ii) conversion of agricultural land, (iii) interference in natural flood drain, (iv) access of water front to other users, (v) soil salinisation, (vi) salinisation of drinking water, and (vii) nutrient loading and self pollution.

The emergence of these issues are basically due to the poor siting of the farms. Presently clear guidelines exist and the Coastal Aquaculture Authority (CAA) does not permit shrimp farms to be constructed in mangrove areas, agricultural lands, saltpans and other commercially utilised lands. Further provision of buffer zones between aqua farm to farm, farm to agricultural land, farm to villages and farm to other ecologically sensitive areas is also being made mandatory for obtaining license from CAA. The extent of buffer zone required will depend on the site characteristics, soil quality and tidal conditions. Construction of shrimp farms in seepage areas is not permitted so as to avoid the problems of salinisation. The number of farms that can be constructed in a given area should be decided after carefully considering the carrying capacity or assimilation capacity of the water source. It has been mandatory that larger farms of above 40 hectares should conduct an Environment Impact Assessment study and also have a Environment Monitoring Programme which are specific to the particular site.

The adoption of a proper farm design will reduce the problems of water quality encountered by the farmers. The design characteristics should be very site-specific and while designing a farm, the tidal characteristics, the soil quality and the source of water quality should be taken into account. In areas, where the source of water is highly turbid with suspended particles, an intake reservoir for settling the silt is very essential. Similarly, in areas, where there is an overcrowding of farms and the intake and outfall are from the same creek, the intake reservoir with provision for treatment of water is essential. In areas where the tidal current is swift and tidal amplitude is high, the waste water from the farm can be directly let out during the low-tide. But in areas where the tidal current is very low, it is essential that the wastewater is treated in Effluent Treatment Plant (ETP) before release into the natural system. Though ETP has become mandatory for larger farms of above 5 ha, it is necessary that smaller farms that are located in close proximity to each other should also consider and adopt the setting up of common ETP to avoid self-pollution. The design of the farm should not in any way prevent the natural (flood drain from the village) water flow.

Farm construction practices play an important role in the management of environmental interactions, such as, salt water intrusion caused by seepage from ponds can be easily controlled by careful compaction during dyke construction and by siting farms on clay soils.

Pond preparation by way of drying, ploughing and liming are required to correct the pond soil conditions with regard to the accumulated nutrients and acidity. In high density farming system, there is a possibility of the pond sediment becoming unusable during the next farming cycle due to high accumulation of nutrient loads. Disposal of such sediments outside will be an environmental hazard. But in case of low density farming practices, such problems are rarely encountered.

Development of natural feed, through enhancing nutrient levels through fertilization, is very much required, since hatchery reared seed require natural feed during the initial few days of rearing. Care should be taken to avoid algal blooms in the pond which will lead to oxygen depletion. Application of fertilizers in ponds should be regulated depending on the phytoplankton density. Water transparency expressed as Secchi disc visibility can be taken as an index of phytoplankton abundance.

Collection of wild post larvae for stocking in the farm during early 1990s had serious implications on the biodiversity of the natural ecosystem. With the establishment of more than 300 hatcheries during subsequent years, the dependence on wild seed has come down. Now it is mandatory that all the shrimp farmers should stock only hatchery reared seed, and most of the farmers undertake PCR testing of the seed to ensure virus free seed.

The disease outbreaks in shrimp farms have made the farmers to realize that lower stocking densities lead to sustainable harvests. For *Penaeus monodon,* the stocking density upto 10 nos per m^2 will be ideal. In hatcheries the post larvae are reared in salinities above 30 ppt, but in the farms the salinities vary considerably. Hence it is essential that the shrimp seed is acclimatised gradually to the pond salinity before release. This will reduce the stress of the stocked shrimps.

Nutrients and organic wastes produced in shrimp ponds consist of solid matter (mainly uneaten feed, faecal matter and phytoplankton) and dissolved metabolites (mainly ammonia, phosphate, carbondioxide, nitrite and nitrate). Various management methods are followed to maintain these within the tolerance limits. Among these, the cheapest one is of water exchange. Shrimp farmers adopt a water exchange of 5-30 per cent per day depending on the availability of water and the quality of pond water. In low density farming, high level of water exchange is not required. In view of the complaints of nutrient loading in the open environment and the fear of viral contamination in the source water, the water exchange rates have been reduced by the farmers. Some farmers even follow no-water exchange during the first two months of rearing. In other countries, importance is being given now for developing "zero-water exchange" system with recirculation practice.

Feed quality and conversion ratio have considerable influence on the waste levels. Reduction of phosphorus content in feed, control of dietary nitrogen in relation to metabolism and improvement in physical characteristics, such as, attractability, water stability, texture and appropriate size of the feed will help to reduce the nutrient loading to a large extent.

Feeding rate is very important in shrimp farming. Over feeding leads to feed wastage and spoil the water quality; while the underfeeding affects the growth of the shrimp. One of the ways to regulate feeding rate is through feed check trays, kept in the pond. Feed quantity should be regulated according to the feed consumed in check trays.

Since prawns normally take 4-5 hours for digestion of feed, the total feed may be divided and fed 4 to 5 hour intervals. Shrimps are mainly nocturnal in their feeding habit and hence more quantity of feed is to be provided during the night time. Of the total quantity of feed, 35-40 per cent may be distributed during the day and 60-65 per cent during night. Such a schedule should be regulated through observations of feed consumption from check trays. Feeds with high acceptability, high digestibility and assimilation will reduce waste generation and nutrient loading.

Disease in one of the most important problems in aquaculture. Disease manifestation is an expression of a complex interaction of shrimp, pathogen and environment and it can be controlled only through holistic management approach. Important causative agents of infectious diseases in shrimps are viruses, bacteria, fungi and parasites. Among these, viruses form the serious pathogen of the shrimps. White spot virus is presently causing havoc in the prawn culture sector.

Although judicious use of chemotherapeutics may be required in specific cases, Indian shrimp aquaculture witnessed indiscriminate use of chemicals to control diseases without much success. As there is no drug which can control viral diseases, use of these should be discouraged in view of the existing strict regulations in international market on the residues. MPEDA has banned the use of twenty such drugs in shrimp farming.

It is a fact that control and prevention of diseases in aquaculture is a function of management. Disease problems arising in aquaculture can be attributed primarily to the environmental conditions and most of the pathogens are facultative pathogens in nature. So the management of the pond environment is of utmost importance for disease prevention and control.

Prevention of horizontal transmission of virus through strict bio-security protocol is essential. This includes, tyre-bath, foot bath, hand wash which are required to be provided. Chlorination of intake water in reservoirs, fencing around the farm, crab fencing, bird fencing and use of independent implements and personnel for each pond.

The waste from shrimp farm contains mainly suspended solids (for physical qualities of feed and fertilization levels) and dissolved nutrients (by the chemical composition of the feed ingredients and the fertilizers). Biological and chemical oxygen demands of the waste water are an indication of the level of microbial and chemical interactions. Sedimentation of the wastes is the most cost effective method of treating them. The dissolved nutrients could be removed by farming sea weeds, which can absorb them and the light suspended solids can be removed by culturing molluscs (filter feeder). A standard guide line for shrimp farms in this regard can be available from CAA.

Harvesting of prawn can be done by completely draining the pond and hand picking and trapping. The drained out water for harvesting should be pumped into waste stabilization ponds and kept for a few days for settlement before releasing into the open water. Icing should be done immediately after harvest. Generally the processors or buyers collect the harvest from the farm site in refrigerated vans. When such a facility is not available and the produce has to be transported over a long distance the shrimps should be beheaded and stored in ice to prevent spoilage.

Use of Probiotic for Disease in Shrimp Farming Ponds

One of the major problems at present in shrimp farming is microbial disease. The application of antibiotics has been of no use as a remedy to the problem. The use of probiotics to control pollution as well as disease and for sustainable and eco-friendly aquaculture is gaining importance. Aquaculture probiotics are a scientifically blended concentration of selected, adopted and cultured bacterial formulations plus enzymes and special buffers fermented with cereal and mineral substrates. These are applied to shrimp farm ponds for having disease-free shrimp production.

The beneficial bacteria present in the environmental pond probiotics are *Bacillus* sp., *Pseudomonas* sp., *Nitrosomonas* sp. etc.

The pond bottom conditions get deteriorated by the left over feed, accumulated faecal matter and dead algae. The outbreak of disease is mainly due to deteriorated water and soil conditions. The major functions of probiotics are to degrade organic waste in sludge accumulated at the bottom, and to accelerate the removal of hydrogen sulphide and neutralisation of other harmful substances.

Due to good environmental conditions plankton bloom stabilizes and feed consumption increases. There was improved growth rate and increase in average body weight. Further, disease outbreak in ponds will be prevented. Probiotics are also mixed in feed and fed to prawns. This enable beneficial bacteria to proliferate in the digestive tract, and lead to production of enzymes. These help in the digestion of feed in digestive tract.

Emiron AC from Biostadt India Ltd, a probiotic was used in shrimp pond situated in Chandipur area, Balasore coast, Orissa for improved environmental conditions. The applied dosages were 10 kg/ha at every 10 days interval. The pond of 4000 square metre with a stockings density of 10 prawns (*Penaeus monodon*) per square metre was taken for the study together with an identical size and condition as control. Water samples were collected periodically from soil-water interface and subjected to microbial analysis for entropathogenic bacteria.

In the ponds where probiotics were used, there was a gradual increase in *Vibrio* yellow and *Vibrio* green colonies from the day of stocking till the harvest. Shrimp production in this pond was 1528.36 kg with an average body weight of 34.8 gram. No incidence of disease was observed. Where as in the pond without probiotic treatment *Vibrio* green and yellow colonies reached high levels (240 and 650 CFU/ml respectively) and outbreak of Vibriosis disease took place. Secondary infections like black gill disease, protozoan infestations caused by *Zoothamnium* and fungal infections were observed. As a result distress of prawns lead to early harvest of prawns with average body weight of 16.51 gram. The total quantity of prawn harvested from the disease affected prawns was 365.13 kg.

Prawn Farming Practices in Goa

There are around 45 shrimp farms in Goa, covering an area

about 150 ha. Most of the farms are creek based and tiger prawn, *Penaeus monodon* is the only species farmed following the extensive farming system.

The pond bottom is dominated with sandy clay soil (52.9 per cent) followed by clayey loam (35.3 per cent) and silty clay (11.8 per cent). The pond bottom is sun-dried and ploughed using either tractors (54.5 per cent) or bullocks (45.5 per cent) after every three or four crops. High seepage problem and crab menance, particularly on the lower parts of the dykes, where the soil is sandy clay in nature. Liming of the pond is done using lime stone (39.9 per cent) dolomite (33.3 per cent) and quick lime (19.9 per cent), while some farmers use all the three types of lime. A wider variation in lime application, ranging from 200 to 2000 kg/ha is observed due to the varying acidity of the soil to neutralise the soil pH to the optimum level.

Cattle dung, and fertilizers, such as urea, single supper phosphate, triple super phosphate and diammonium phosphate, are added to the pond water after its first filling, with a gap of five to ten days from the day of liming. The fertilizers are applied at the rate of 20-70 kg/ha with an average quantity of 35-38 kg/ha during the crop. The plankton blooms are observed in most of the farms in the subsequent culture period. These are observed to be moderate, imparting a typical green colour to the water, suitable for rearing shrimps.

All the shrimp farmers stock their ponds with the seed procured exclusively from hatcheries either from Tamil Nadu (80 per cent) or from Karnataka (20 per cent) states. Before stocking, the seed (PL15–PL 18) is tested for pathogens, particularly for the White Spot Syndrome Virus (WSSV) and Monodon Baculo Virus (MBV); and is stocked in the month of March for the first crop and in September for the second crop. The stocking rate varies 5 to 15 post-larvae/m^2.

Commercial feeds, of various types, namely, the starter, the pre-grower, the grower and the finisher along with the grades in between them, are used by the shrimp farmers. The feed is generally broadcast by hand from the dyke to the peripheral areas of the pond. The frequency of feeding varies from four to six times a day. In four-times-a-day feeding schedule, the feed is distributed around 06 hr, 12 hr, 18 hr and 24 hr successively. The changes in feeding time and frequency are made according to the situation, such as, excess feed left over in the check trays etc. The total feed used per hectare per crop of about four months ranges from 1.5 to 4.5 tonnes with an average of 2.74 tonnes and the feed conversion ratio (FCR) is reported to vary from 1:1.4 to 1:1.6. The majority of the farmers (58.5 per cent) have reported FCR of 1 : 1.5, while 23.3 per cent farmers reported on FCR of 1 : 1.6 and 18.8 per cent of them as 1:1.4.

The pond water exchange is made in a combined manner using both natural tidal ebb/flow currents through the conventional system of operation of sluice gates and pumping in the water. A maximum exchange of 20-30 per cent is followed, each time, by about 50 per cent of the surveyed farms. Some of the farms (25 per cent) exchange only 5-10 per cent water, while 18.8 per cent of the farms exchange 10-20 per cent every time. In the remaining 6.2 per cent of the farms zero water exchange is practised with application of probiotics to maintain the water quality throughout the culture period. Although this adds to the cost, entry of infected water from the common source is effectively prevented. Nearly 90 per cent of the farms have the water inlet and outlet structures on the dykes in the opposite sides of the ponds, while the rest of the farms, which are of irregular shapes, have the inlets and outlets at the same side.

Each farm uses 1-7 pump sets depending on the financial capacity of the farmer. Two large farms of 5 ha and 5.5 ha use 7 and 6 pump sets, respectively, while other small farms of less than 3 ha use two or three pump sets each. Depending upon the farm size, on an average 4-6 aerators per hectare and used for aeration of the pond water. About one third of the shrimp farms, irrespective of farm size, have one generator per farm and rest of them use public electric supply.

Periodical monitoring of soil and water quality, health condition of shrimp, feed consumption and growth parameters are followed in all the shrimp farms. While physico-chemical parameters are measured once in 2 or 3 days, the general appearance of the prawn for any disease attack and feed consumption is observed daily. Sampling with cast net is done to monitor growth once in 7 to 10 days. Comprehensive checking of the stock is done fortnightly (in 50 per cent farms), weekly (in 15 per cent). The sample size taken for the shrimp health and growth checking is less than 5 per cent of the total stocked animals.

Screening of the seed for viral diseases, such as WSSV and MBV is carried out as a routine practice in the hatcheries using commercial PCR test kits. In the course of the crop period, observations are made on the body colour, condition of appendages, colour of the gills, fullness of the gut, stiffness of the shrimp and its grip to hold when kept on the finger of the sampling person, and presence and absence of black and white spots.

Improvement of pond water condition, after first filling is done using bleaching power. Application of tea-seed oil cake, ammonium sulphate etc is also practised for eradication of pests in the pond before stocking the shrimp seed. A variety of aquaculture product (Iodine, lime, netsol, sanmoult, zeolite, addoxy, vitamins, aqua-pro, aqua-c, vitamin-c) are used by the farmers for the general health as well as for pond bottom hygiene, better moulting and growth of prawns. Probiotics (Environ AC, Super PS, Wunnpu, Thionil, Epicin, Aqualact, Super biotic, Ecoforce Baciplus EM-AQ) are used at varying concentrations.

The harvesting is done after 120-150 days of farming, one or two days earlier to full or new moon day and commencing at the start of low tide. The bag nets tied to outlet sluice are used in 90 per cent farm for harvesting and in 10 per cent smaller farm drag net is used for harvesting. Immediately after the catch, crushed ice is used to pack shrimp in plastic tubs and is handed over to the traders, who in turn supply them to processing plants.

The produce is sold exclusively for export market that fetches a price of Rs. 235 to 300 per kg. On an average by investing Rs. 2,47,000 for operational cost (as majority of the farms are owned by the farmers) a return of Rs. 4,60,000 can be realised with a net income of Rs. 2.12,400 per hectare per crop.

Monosex: All Male-Prawn Farming

Heterogeneous individual growth (HIG) is commonly exhibited by male prawns by which some of the males surpass other individuals in size. Various management strategies including improvement of the environmental and nutritional conditions, and manipulating the stock structure by selective stocking and harvesting, claw ablation etc have been effective to some extent in overcoming the problem. But the labour costs involved and the associated problems of repeated harvesting, disturbing the pond bottom, and possible infections affecting prawns after claw ablation etc are the drawbacks. Further, the presence of females in the farming stock would result in energy loss by individuals for reproduction leading to more differential male growth and inhibition of female growth. In view of these constraints, the concept of growing males alone has proved to be a better alternative way for overcoming differential growth and improving the yields. This strategy would also sort out to a great extent the differences in feeding rates of individual prawns and serve to minimize the effects of territorial behaviour of the males.

All male farming of *Macrobrachium rosenbergii* has several advantages over the conventional farming of males and females together, in improving uniform growth rate, feed conversion efficiency survival rate, individual prawn size and overall output from farming. The final mean individual size attained by the prawn in all male farming is better than conventional mixed farming of males and females. All male population precludes breeding activity in farming ponds, thereby being beneficial in diverting reproductive energy of males in the population towards improving growth and yield. The FCR obtained in all male farming is also comparatively lower (1.2 : 1) than from mixed farming (1.9–2.0 : 1) which is a critical factor saving feed costs and affecting profitability of the farming operation. Mixed farming usually extends for a longer period (8-10 months) and may involve as many as 4 to 5 harvests during this period. The shorter duration of farming and lower stocking density used in all-male farming imply less of pollution and water quality problems.

The farmers of Andhra Pradesh have played a key-role in popularising the all-male prawn farming system in which males and females are manually segregated soon after nursery rearing. The post larvae which has grown to juveniles (3-5 g size) after nursery rearing period of 45-60 days are sorted out by trained segregators within 60 days. Manual segregation of males and females is carried out by visually examining the juveniles at the area of the base of last pair of walking legs wherein the gap between the origins of the two pairs of walking legs is less in the case of males, and additionally they possess a shield-like covering that hides the gonopore. Skilled segregators (including many women) could easily seperate 3000-5000 juveniles per day in 6-8 hours enabling a fast popularity of this practice. In 2006, almost 60 per cent of the prawn farmers in Nellore district of Andhra Pradesh employed all-male monosex farming by visual segregation.

The success of all-male farming could also be partly attributed to the lower stocking densities used. The segregated male juveniles were stocked at a low density of 12500-15000 juveniles/ha in the grow-out pond. After 90-120 days (depending on the size of prawn) in the growout pond, seining is carried out for the first harvesting to remove all the fully-grown blue-clawed males (70g-100g) and any females (30-40 g) that may be accidentally present. About 20 percent of the large prawns are thereby removed at this stage, while the smaller prawns are returned to the ponds, allowing them to grow further. After 30 days, the second drag netting is undertaken which is the major harvest in which as much as 60 per cent of the full-sized males are caught. Final harvest is carried out after another 30 days to complete the farming period of 6–8 months (including nursery period).

Manual segregation of juveniles, however, have several disadvantages. It involves continued harvests from the nursery ponds exposing the juveniles to stressful ordeals, reduce survival rates in grow out ponds for examining sex and for high expenditure (as high as Rs. 2/seed). Manual sexing may not be feasible or economical always; and, therefore, it is important to seek alternative methods of producing all male populations.

Though all-male prawn production by neofemale technology, where the androgenic gland (AG) was surgically removed from juvenile *M. rosenbergii* at an early development stage for complete sex reversal, leading to the development of functional females capable of mating and producing progeny, and these fertile sex-reversed prawns (neofemales) when crossed with normal males, it resulted in an all male progeny; a suitable biotechnology would have to be developed and a specific androgenic hormone has yet to be identified in decapods to avoid wide range of abnormalities in andrectomised males depending on the age at which the andrectomy was performed. Till then, aquacultural and biotechnological research and development may continue in establishment of monosex farming system through manual segregation, together with the application of selective harvesting and claw-ablation, as well as examination of different monosex farming strategies under a variety of economic conditions.

Contract Farming of Prawn Culture–Possibility

Indian culture shrimp industry, ranging from culture to exports has been encountering serious problems that have the effect of impeding shrimp production. In order to address some of these problems, which are of a critical nature, Chennai based Oceanaa Group has entered into an Memorandum of understanding (MOU) with Indian Overseas Bank Ltd. (IOB) for taking up shrimp culture under contract farming with stakeholders, such as, the farmers and an insurance company.

The company (9 group company) of net worth Rs. 500 crores is well wedged into software development with an employee strength of 250. Its major thrust is into BFSI domain. A few of the products are used by many leading banks. It has developed the traceability software for shrimp hatchery and culture. Particulars of farmed shrimp production by Oceanaa group has shown a growth of 25 per cent in 2005 over 2003. Though, there is a downward slide of 28 per cent during 2006 and 2007, indicating that there is a urgent need for the government (National Fisheries Development Board) to place the shrimp industry back on the rails through subsidy intervention.

Operational Modalities

The scheme of financing shrimp farmers or groups under the MOU will be implemented in the state of Tamilnadu and Union Territory of Puducherry.

Oceanna group will identify eligible farmers who have obtained license from CAA (Coastal Aquaculture Authority) and satisfy itself about the suitability of site by testing soil/water and other parameters.

The Oceanna group will submit the list of eligible farmers or groups to the Bank for sanction of loans. The selected farmers or groups shall be trained by the Oceanna Group in the scientific management of prawn culture. Oceanna Group will supply seed, feed and chemicals to the farmers or groups after receipt of payment from the bank. Oceanna Group will provide free technical consultancy for pond management and will facilitate insurance of culture ponds and animals. The Oceanna Group will make necessary arrangement for procurement of harvested shrimps based on count and quality. Oceanaa Group will provide farm record registers to be maintained by the farmers or groups which will be supervised by staff of the company, to evaluate survival, FCR and body weight. The Oceanaa Group will remit sale proceeds to the bank for credit to the farmers accounts with the concerned branches of the bank within 7 days of procurements of harvested shrimps. The company will facilitate insurance claims in case of any eventuality in the culture ponds or shrimps.

The scheme of financing the shrimp farmers or groups for shrimp culture under contract farming will be implemented through various branches under the administrative control of Indian Overseas Bank at various places in the state of TamilNadu and the UT of Puducherry.

On final selection of shrimp farmers or groups for shrimp culture, the bank will initiate following steps.

To meet the genuine needs, the bank will prepare unit cost to be adopted, which will be reviewed, once in a year or two years as the case may be. The loan amount relating to working capital requirements such as, seed, feed and chemicals shall be paid directly to the supplier against invoice or receipts. The loan amount relating to electricity, labour, miscellaneous expenses will be directly paid to farmer or groups. The premium for insurance of shrimps will be debited to farmer or group's account and paid directly to the insurance company. The bank shall credit the sale proceeds received from Oceanaa to the borrowers loan account and the surplus, if any, shall be credited to SB account of the borrower kept at the respective branches.

To avoid multiple financing the branches of the Indian Overseas Bank will take necessary steps as per the Bank's laid down instructions. Margin money will be 25 per cent of the total project cost or as per bank norms. Interest rates as applicable to Agriculture advances will be adopted. Bank's security norms as applicable to the facility will be adopted.

Cash credit can be renewed after repayment of bank loan. Prices of prawn indicated in the project are based on the previous years average farm gate prices. However, the payment for the current year will be based on prevailing market prices at the time of harvest.

Insurance of shrimp at the rate of 2 per cent of the total input cost (now 4 per cent of the total input cost). Culture period may be extended upto 20 weeks. The insurance premium rate is 4 per cent of the input cost, which is taken at Rs. 2,50,000/ha water spread area. Though premium rate is comparable with industry standards, its works to Rs. 11,300/ha. In absolute terms it is very high for farmers.

Under the above circumstances NFDB can play a significant role through interventions, such as grants, subsidy etc. Animal Husbandry and diary sector are supported by Department of Animal Husbandry and Diarying by providing subsidy. Some of the subsidy interventions that can be considered by NFDB are mentioned below;

1. Subsidy for shrimp feed at the rate of 10-50 per cent depending upon the farm size.
2. Insurance premium subsidy at the rate of 50 per cent of previous amount.
3. Interest subsidy for interest above 7 per cent per annum.

Table 22: Economics of Shrimp Culture Under Contract Farming

1.	Area–1 Ha water spread area	
2.	Culture period–133 days (19 weeks)	
3.	Stocking density–10 PLs per square metre	
4.	Total number of Pls required–1,00,000	
5.	Cost of seed at the rate of Rs. 0.25 per PL–Rs. 25,000	
6.	Survival–75 per cent	
7.	Average body weight per shrimp–30 gm.	
8.	Total gross weight–2250 kg	
9.	Cumulative feed consumption–3500 kg	
10.	F.C.R (Feed conversion ratio)–1.5	
11.	Feed cost at the rate of Rs. 50/kg–	Rs. 1,75,000
12.	Pond preparation charges–	Rs. 5000
13.	Cost of chemicals, Probiotics etc	Rs. 15000
14.	Labour charges	Rs. 30000
15.	Electricity charges	Rs. 50000
16.	Harvesting charges	Rs. 3000
17.	Insurance of shrimps	Rs. 6060
18.	Total expenses	Rs. 3,09,060
	Rounded off	Rs. 3,10,000

Contd...

Table 22–Contd...

19.	Less 25 per cent margin	Rs.	77000
20.	Bank loan	Rs.	2,33,000
21.	Income from sale of 30 count shrimps at the rate of Rs. 275/kg (for 2250 kg).	Rs.	6,18,750
	Repayment of bank loan		
(a)	Principal	Rs.	2,33,000
(b)	Interest at the rate of 12 per cent per annum for 5 months	Rs.	11,700
22.	Total repayment	Rs.	2,44,700
23.	Net income (21-22)	Rs.	3,74,050
24.	Net surplus (21-18)	Rs.	3,09,690

Subsidy support interventions will go a long way in rejunuvating the shrimp industry. Small and big farmers who have interest in this activity will have to be encouraged to sustain the activity on a long term basis adopting best management practices.

Constraints in Adoption of Improved Shrimp Farming by the Farmers

Data were collected in respect of operations at 40 shrimp farms of south Konkan region so as to utilise the same to establish relationship between various personal variables and technology adoption behaviour of shrimp farmers of south Konkan region and to understand the constraints faced by the farmers in the adoption of improved shrimp farming techniques.

Adoption of any innovation depends upon socio-economic status of farmers. Personal variables of a social nature play an important part in the adoption of improved techniques by the farmers. Sustainable and improved shrimp farming practices have the potential to attract fresh entrepreneurs. These will also facilitate existing farmers to improve their per unit production. In orders to achieve eco-friendly and sustainable development of prawn farming, certain interventions including adoption of appropriate policies and planning and improved practices are necessary. These interventions would have to be targeted especially towards the promotion of an equitable distribution of benefits among stakeholders concerned and also towards the long term environmental sustainability of prawn

Table 23: Constraints Faced by Shrimp Farmers of South Konkan, Maharashtra

Sl.No.	Constraints in adoption	Districts		Total
		Ratnagiri	Sindhudrug	
1.	Unavailability of disease diagnostic laboratory	11 (68.75 per cent)	21 (87.50 per cent)	32 (80.00 per cent)
2.	Unavailability of quality inputs in local market	11 (68.75)	20 (83.33 per cent)	31 (77.50 per cent)
3.	Exploitation of farmers by middlemen	8 (50.00 per cent)	14 (58.33 per cent)	22 (55.00 per cent)
4.	Lack of proper extension network	8 (41.67 per cent)	10 (45.00 per cent)	18
5.	Lack of technical knowledge	8 (16.67 per cent)	14 (30.00 per cent)	12
6.	Unavailability of PCR–testing laboratory in the vicinity	5 (31.25 per cent)	5 (20.83 per cent)	10 (25.00 per cent)
7.	Unavailability of soil-water testing facility	4 (25.00 per cent)	5 (20.83 per cent)	9 (22.50 per cent)
8.	Unavailability of technical persons for farm construction	4 (25.00 per cent)	4 (16.67 per cent)	8 (20.00 per cent)
9.	Inadequate finance	3 (18.75 per cent)	3 (12.50 per cent)	6 (15.00 per cent)

farming. It is also quite essential to record and monitor data for the effective adoption of improved techniques by the farmers, taking care for the elimination of constraints in the process. The adoption of most of the technologies recommended for prawn farming by the targeted end users is generally weak. A study of constraints faced by shrimp farmers in the adoption of improved techniques and the formulation of measures for neutralising them will upgrade the quality of the results of the final outcome of the effort at improved farming results (Table 23).

Most of the farmers of south Konkan region (80 per cent) faced the problem of disease diagnostic laboratory in the vicinity. Setting of disease diagnostic laboratory is beyond the capacity of marginal farmers due to high investment. Government agency can set up such laboratory to serve the needs of shrimp farmers. Unavailability of quality inputs locally led the farmers to enhanced variable cost due to additional payment of middlemen and transport. Establishment of shrimp farmers cooperative store may help to solve the problem of getting quality inputs at lesser price. The farmers were unaware about the availability of soil and water testing laboratory in both the district (Ratnagiri and Sindhudurg). Farmers did not face the problem of PCR testing of seed prior to stocking, because PCR testing was mainly done at hatchery site and marginal farmers procured seeds through middlemen or along with consignments of seed procured by other farmers. Collective efforts are needed to solve the problems faced by the shrimp farmers to promote the shrimp farming activity on a larger scale in the area of difficulties.

Chapter 13

Posteriharvest Handling and Transportation

Distribution

Prawn lose freshness quickly after they are caught and killed, and it is essential to maintain the freshness of catches on board until they are landed. Also care has to be taken of the freshness throughout all of the stages of handling and transportation until they are eventually cooked and eaten by consumers in cities and towns. The basic conditions that facilitate the distribution for prawn fishery in a country, a system of prompt and smooth transportation with careful maintenance of the freshness is of first importance. Next is to set a reasonable price for them. The third is that the payment of all the services and the goods has to be done fairly and promptly. Last but not the least, is to reduce the cost of intermediary traders by simplifying the distribution system as much as possible.

In India, the distribution system of prawn fishery from fishermen, through various intermediate stages to the final consumers is diagramed in Figure 64.

As far as the coastal fishery is concerned, all the prawn catches are landed at landing centres, which are located all over the country. The catch is sorted by the kind and size of prawn, auctioned, and then through the intermediary traders they are forwarded to the

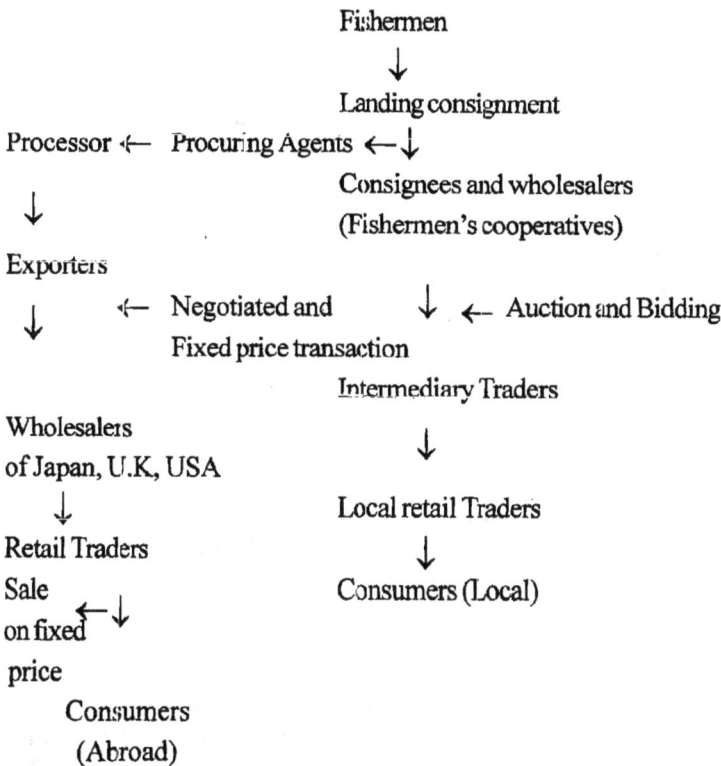

Fishermen

↓

Landing consignment

Processor ←— Procuring Agents ←—↓

Consignees and wholesalers
(Fishermen's cooperatives)

↓

Exporters

↓   ←— Negotiated and   ↓ ←— Auction and Bidding
Fixed price transaction

Intermediary Traders

Wholesalers
of Japan, U.K, USA   ↓

↓   Local retail Traders

Retail Traders   ↓

Sale   Consumers (Local)
on fixed ←—↓
price

Consumers
(Abroad)

Figure 64

markets in the consuming cities, processing factories and local retailers.

Thus the consuming markets are able to gather small consignment of prawns from various local landing centres after meeting the requirement of processing plants. The sorting, auctioning and allotment are done and they are again sent out to where they are eventually in demand. To an unfamiliar eye, the system may seem to be complicated, but the advantages of having markets in both landing centres and consuming centres functioning in cooperation with each other may be as follows; (a) At the landing centres catches are sorted depending on the kinds of prawn, size and quality and then forwarded to the consuming markets surely and promptly according to difference in the usage, such as, use as fresh, processing and subsequent export and so on. (b) Owing to the facts that all types of prawns are gathered all together, though in small quantities, in the local production, they can meet the various demands of the consumers easily. (c) The quality of the products is always checked at every stage between the markets and landing centres, thus it makes the price correspond to the quality.

Since the 1960s some progress has been made in the facilities for transport of prawn, especially of export quality, because of the development of techniques for cold and frozen storage and the improvement of distributing systems. Today in India, many of the activities of producing, processing and exporting prawns are united into a network.

The Catch

In case of coastal fishing, the prawn catch are kept with or without ice until landing, thus they are handled as fresh. In case of off shore and deep sea fishing it takes many days to reach the fishing grounds and return to the home ports and also the boats are large in size. All the boats of high-sea fishery and some boats of off-sea fishery are equipped with freezing facilities, which bring back the prawn to ports, while keeping the original freshness. Small boats, fishing in coastal waters carry ice boxes to keep the prawns as soon as they are caught to avoid losing freshness. The catches are quickly transported by insulated vans, trains or boats to all the areas of the country. The catches are packed in cartons, chilled with ice or frozen. Transportation of prawn without ice is rarely seen today.

Maintaining the Freshness

The fundamental techniques required for products of good quality are to do some pretreatment of prawn to prevent autolysis and putrefaction. It should not be forgotten that these techniques allow smooth transportation and marketing for the fishery industry.

Starting from the moment when the prawn is caught on board, then transported over land, till it is eaten at home. There are many details to be learned scientifically about the prawn meat itself and its condition during handling.

On Board

Even in the case of small fishing boats which expect only a few days voyage, care must be taken to keep the catch in good condition.

There are a variety of ways of handling depending on the kind of prawn, size and fishing method.

1. The prawn may be kept alive in a living well on the boat. Sea water should be kept circulated through the well.

2. The scupper of the prawn preserve on board may be closed. It should be filled with ice and the catch.

3. After sorting outs the prawn, they are to be put in cartons, the gaps to be filled with ice and then the cartons may be kept in the hold. The wall of the hold should be of heat arresting construction. When the catching activity is prolonged, the catch has to be put in the hold in the order they are caught inorder to maintain the quality of the whole.

In case of middle and large sized boats, provided with ice storage, prawns are to be sorted out, packed them in cartons with crushed ice and then placed them in insulated hold. The cold storing boats are equipped with freezing machines on board but not a quick freezing system. The holds are pipelined on the wall with freezing medium and the ice packed prawn cartons are placed inside. The prawn may also be put in bulk, that is, without packaging. Freezing boats, are equipped with quick freezing machines. The prawn are put in cartons and frozen before being put in the hold.

On Land

An essential part of the distribution system for fishery products is to maintain the original freshness of fresh prawn, frozen prawn and processed products by means of ice, cold and frozen storage. In 1884 when ice factory was established for the first time, the techniques in keeping fresh foods have constantly been developing. In the 1920s a freezing system was adopted on board off-sea fishing boats for freezing the catch. Ice making and storing, freezing and refrigerating are inseparable technological systems. The progress of technology in these areas has contributed greatly to enlarge the fishing grounds of off-sea and high sea fisheries and to modernization in distributing the fishery products through networks. This in turn has made possible a many sided development of prawn processing industry and brought about many new factories over a growing area.

Methods of Ice and Various Storage

Ways to storage at low temperatures can be classified into the following three methods.

Ice Storage

Keeping the prawns either in crushed ice or ice water.

The sarcoplasmic protein nitrogen decreased only by 3-6 per cent during about 2 weeks of storage in ice in case of two species of marine prawns (*Penaeus indicus* and *Metapenaeus dobsoni*). Myofibrillar protein nitrogen, on the other hand, showed a decrease of 15-20 per cent during the same period in prawns, which tends to indicate that this protein gets denatured with storage time. The decrease of myofibrillar protein nitrogen is accompanied by the increase of denatured protein to the extent of 20-25 per cent during the same period. There is only slight variation in the amounts of stroma protein during ice storage of prawns.

Cold Storage

Using a refrigerator the prawns are kept at low temperatures but barely frozen (–1°C to +1°C).

Frozen Storage

By freezing machine the prawns are frozen at about–10°C, or below–20°C, when quick freezing is needed. Then the prawns are kept at low and constant temperatures. Ice and cold storage can preserve the prawn at rather low cost. They are however, inadequate to stop the actions of autolysis and bacteria, thus prawn gradually deteriorate during storage. The limit for keeping the original quality may be ten days for ice storage and one month for cold storage.

Freezing storage is carried out now a days at very low temperatures. The prawn freezes quickly and it keeps its original quality for a very long time. To restore the original quality before cooking frozen prawn, thawing has to be done gradually. If duly treated, the prawn will be as fresh as just-caught one. Any change in the quality during frozen storage is so slight that the prawn can be stored for even longer period than a year.

Dormant Live Prawns Flown Daily: An Advantage to Culture Fisheries

Air-shipment of live prawns is a unique practice and is supported by prawn culture in man-made circular tanks and high level of consumption in the large cities. Producers airship live prawn to a distance of about 1000 km away.

Prawns are immersed in refrigerated water for two to three hours on the morning of the day they are to be shipped as a means to retard their normal physiological functions. The prawns are then packed in cardboard boxes filled with sawdust and flown hundreds of kilometers to the urban centres. On arrival they are promptly released in the water tanks to resume life process.

Storage in Cold Temperatures

In every fishing port, transit station and in consuming cities large cold storage are equipped with freezing machines. The purpose of having a cold storing system is to keep the original freshness of the prawn and also to control supply to the markets. At the landing centres, when boats have too big a catch and they rush to small port without enough capacity in transportation facilities, it has to be cold stored for a while. Further, when there is a period of large landings by which a considerable fall of price in the consuming market is expected, cold storage is used while trying various means to maintain a reasonable price. At the consuming centres and transit stations the consignments are temporarily kept in cold storage awaiting the shipment for export, the auction activity and transportation and also to control the amount of supply to the markets. At either the landing centres or the consuming markets, the processing factories may cold store the prawns until the time it is needed. The same is true with case of prawn for local consumption. They are purchased when the price is low and cold stored until sold in retail markets.

Consignment Distribution in Landing Centres

There are about 1500 fish landing centres along the coast of India with markets in each of them. Some are large and others are small. The catches of offshore and deep sea fisheries are landed in large landing centres, while the catches by coastal fishery are landed in small landing centres along the coast. In these landing centres the catches are auctioned, and then transported to consuming markets and processing factories. In case of cultured prawns and some frozen prawn from deep-sea fisheries, the auction at the markets may be skipped and they are transported directly to the processing plants and consuming markets.

Prawn Processing Industries

In localities where prawn is abundantly available and is suited for processing, many processing industries (399 with daily freezing capacity of 7283 tonnes) have been developed. Further, in the case of more profitable export industry, such as, freezing of high quality prawn, frozen precooked prawn, the factories are built around the suburbs of cities.

Functions of Fish Markets

In Kolkata, Mumbai, Chennai, Cochin and Vishakhapatnam and other main cities in India there are public fish markets, where the municipalities are responsible for their administration. Here commission wholesalers, who are entrusted by the consignors in the landings ports, participate in the auction. The consignments are purchased by the intermediary traders in the market, who forward the goods in all directions to where they are needed. The trusted and registered merchants auction the goods under a fair competition system, thus reasonable prices of the goods are maintained and safe and prompt payment is carried out as well. In small fishing village, where wholesalers are absent, a fishermen's cooperative will fulfil this duty.

Containers to Transport Prawn

Bamboo baskets have been used for this purpose for many years. In recent years, the prawns meant for export, are transported in plastic boxes and carton boxes for frozen prawn have been in popular use. There is, however, a problem to standardizes their sizes and shapes to improve the efficiency of labour and transportation.

Retailers and Final Consumers

Prawn retailers in India have private shops in the public fish markets. On cemented platform they display the prawns of different kind and sizes, mostly shell on, in the ambients temperatures until the customers buy them. For export in large cities, all the prawns are frozen and packed according to the grade for shipment and export. Prawns are mostly eaten at home, but there are restaurants specializing in prawn preparation.

Prawn Processing Methods

The species of prawn caught in seas are so rich in variety and people's tastes in food are also so varied that the development of shell fish processing techniques has been accelerated in all branches of the industry and a great variety of processed products have become available in the market. Due to the development of cold storing facilities intended for industrial and household use, it has became easy to keep various fishery products in good condition. Processing techniques, such as, drying, salting, smoking etc were originally aimed to keep fish and shellfish edible for long periods without using ice, but the recent development of cold storing facilities has made it possible to add better taste to the product.

Processing for Preservation

A. Drying

Prawn is dried in the sun or over fire to lower the moisture so that the growth of bacteria can be checked.

1. Simple drying–A small quantity of salt is sprinkled on fresh prawn, round or shell removed. Then it is dried in the sun or in hot air from a fire.

2. Salting and drying–Prawn is kept in salt water for 30 to 60 minutes. Then it is dried in the sun or in hot air from a fire. In this case, however, drying is not so good as in (1).

3. Boiling and drying–The method is applied to small–sized prawn. The prawn are boiled in salt water and then dried.

4. Dried pieces of prawn meat–Middle sized prawn are steamed and the head and shell are removed. The block of meat is generally dried over a wood fire while taking care not to raise the temperature of the meat too high. A sophisticated technique of molding on the surface of the meat is included in the process.

B. Salting

Prawn is stored in either salt water of high concentration or crystal salt. The salt penetrates into the prawn body to dehydrate the meat so that germs will not grow.

C. Smoking

Shell fish is dried in the smoke of a wood fire. The heat dehydrates the prawn meat. Some components of the smoke also act against the growth of germs. Processing techniques are aimed at preservation as well as adding taste to the product.

Processing for Preservation and Taste

Small shellfish pieces, sea weeds etc are cooked with seasoning syrup made of soy sauce, sugar etc, until the whole material dries up. The product is kept edible without cold storage.

Half dried prawn products are widely used as snacks with alcoholic drinks. Many kind of shellfish are processed into this category of food.

Either raw prawn or half cooked prawn is put in a can, vacaum sealed and cooked. Brine or oil or some seasoning may be added. Bottling is a simpler processing system but does not keep the quality of the products so long as canning.

Frozen precooked prawn are vacaum packed and freeze stored for household use. They may be coated with bread flour to be fried in the kitchen.

Among value added products, most prominent and lucrative one is the coated product. Quality of coated product generally depends on appearance, colour, crispness, adhesion, flavour, pickup, cooking time, method of cooking and microbial load. Freshly caught shrimp (*P. indicus*) of 16-20 size count was used for coated shrimp products of batter mix (maida, cornflour, bengal gram powder, salt, sodium tre–polyphoshate, turmeric power and hydrocolloid). The product is frozen at–40°C and stored in polyethylene pouches at–20°C.

Fisher women often sell small sized prawns in adjacent localities of the landing centres. Fishermen or their cooperatives sometimes sells small sized prawn in retail shop in cities. These are, however, rather exceptional cases at present. Their aim is to simplify the complicated distribution system with a hope of reducing the intermediary traders margins, lowering the end prices and encouraging the increase of the demand. All of these are expected to result in a better income for the producers.

In the case of cultured shrimp, which are high priced when sold alive. The culturists want to transport them to the processors more promptly. In most cases the processors have their own route of transportation and marketing. The processors mostly export the prawns after processing.

Purpose of Prawn Processing

Differing greatly from agricultural and livestock products, fishery products are unpredictable in regards to quantity of production, and difficult to keep in their original quality.

1. Fishing is usually restricted to a certain season. This is especially true with such abundant prawn migration in coastal seas. In the case of agriculture and animal husbandry, it is easier to control the seasonal fluctuation in the quantity of production, but it is very difficult for the prawn fishing industry with the exception of cultured prawn.

2. The meat of shellfish contains more water and less fat than animal meat and a rapid autolysis takes place soon after death. Infection by bacteria is also rapid.

3. Fishermen's villages and fishing ports are usually localized on the coast of the country as they have to meet the migration of shell fish and other essential conditions for their fishing activities. If they do not have any facilities for forwarding their catch to other consuming centres, then, they will consume only what they like and the rest will go to waste.

Thus it is easily understood that the following activities are important to make the best use of the prawn catch as well as to make the industry more profitable.

1. Immediately after the catch is placed on board or landed, some suitable pretreatments have to be done to keep the freshness and to prevent putrefaction.

2. To balance the production and consumption, some facilities of cold or frozen storage have to be available in the producing centres in order that supply to the markets can be controlled over long or short periods, as is the case with processing for preservation.

3. It is essential to establish a distribution system to transports the products to inland districts while maintaining their freshness.

4. Prepare various processing facilities to improve the quality of products, thus increasing the demand through offering more appealing products for the consumers.

5. It is necessary to encourage the production to prawn fishery products and to devise ways to use unutilized resources.

Kinds of Prawn Processing

Indians are fond of eating fresh prawn. Thus the largest part of the catch is marketed in its fresh state. Since 1950s, with the increasing demand in export market, the percentage of frozen prawn and processed products have been increasing yearly.

Especially with some kinds of frozen blocks of prawn and individual quick frozen prawn (IQF) are shipped to importing countries, like Japan, U.K and USA, where the frozen prawn are brought to retail shops, where they are thawed and sold to the customers, who eat them in their home as raw fish. Techniques of cold and frozen storage are important parts of the fishery industry and they are expected to become all the more important in future.

Dried and salted prawns, especially smaller ones, used to be produced as a side job by fishermen. In recent years, however more industrialized systems of processing fishery products have been initiated to produce prawn paste products, canned products and reconstructed prawn products from small broken prawns.

Latest Development of High Yield of Tiger Prawn

Very high production of *Penaeus monodon* as high as 17.5 tonnes per hectare, on an average, has been reported in Australia. The maximum production of 24.2 tonnes per hectare was observed at the Australian shrimp farms of CSIRO. The high production was because of a monodon brood stock, which was developed after eight generations of selective breeding, CSIRO's scientists have bred the black tiger prawn that is producing record yields. One of CSIRO's partners, Gold Coast Marine Aquaculture in 2010 produced average yields of 17.5 tonnes per hectare, more than double the industry's average. Several ponds produces 20 tonnes per hectare and one pond produced a record yield of 24.2 tonnes per hectare. Dr. Nigel Preston, the leader of the project, claimed that these huge yields can be replicated year after year, which means consistent supply of a reliable and high quality product–all vital factors for the long term growth and prosperity of the Australian prawn farming industry.

References

Ahmad Ali, Syed. Effect of carbohydrate (starch) level in purified diets on the growth of *Penaeus indicus*, Indian Journal of Fisheries, Vol. 29., Nos 1 and 2, 1982.

Ammini, P. L., Lata L. Khambadkar, Sindhu K. Augustine and Sukhdev Bar. A brief report on the marine fisheries of Puri, Orissa, Marine Fisheries Information Service, T and E, Ser. No. 195, 2008.

Aravindakshan, M. and J. P. Karbhari. Some aspects of the fishery and biology of periscope shrimp from Bombay waters, Marine Information Service, No. 83, CMFRI, Cochin, 1988.

Aravindakshan, M; J. P. Karbhari, C. J. Josekutty and J. R. Dias. On the occurrence of *Mesopodopsis orientalis*, Tattersal, a mysid off Maharashtra coast with a note on its fishery, Marine Information Service, No. 81, CMFRI, Cochin, 1988.

Ayyappan, S; N. Kalaimani and A. G. Ponniah. Disease status in Indian shrimp aquaculture and research efforts for better health management, Fishing Chimes, Vol. 29, No. 1, 2009.

Biswas, K. P. . On the behaviour of marine crustaceans in an electric field of alternating current, Fishery Technol., Vol 8. No. 1, 1971.

Biswas, K. P. Inshore demersal fisheries off Orissa coast, Fish Technol, Vol. 16, 1979.

Butler T. H. The Commercial shrimps of British Columbia, Fisheries Research Board of Canada, Report No. 83, 1950.

Chand, B. K. and J. Pol. A new horizon in prawn farming, Fishing Chimes, Vol. 13, No. 10, 1994.

Chandrasekaran, V. S., B. Shanthi, M. Kumaran, and M. Krishnan. Shrimp farming practices in Goa, Fishing Chimes, Vol. 29, No. 8, Nov. 2009.

Chandrasekharan and A. P. Sharma. Shrimp farming and health hazards, Fishing Chimes, Vol. 13, No. 10, 1994.

Chiu, Y. N., L. M. Sandos and R. L. Juleano. Technical consideration for the management and operation of intensive prawn farm.

Corre, V. L. (Jr.). Prawn feeding management.

Devadasan, K. and M. Rajendranathan Nair. Observations on changes in the major protein nitrogen fraction of prawns during ice storage, Fishery Tech., Vol 7, No 2, 1970.

Francis, O. Henry. Shrimp culture under contract farming, Oceanaa shows the way, Fishing Chimes, Vol. 28, Nos. 10/11, Jan and Feb, 2009.

Geethalakshmi, V., Nikita Gopal and G. R. Unnithan. Analysis of Indian shrimp exports and its prices in major international markets, Fishery Technology, Vol. 47 (1), 2010.

George, Ninan, A.C. Joseph, A. A. Zynudheen, A. R. Abbas and C. N. Ravishankar. Effect of hydrocolloids as an ingredient of batter mix on the biochemical, physical and sensory properties of frozen stored coated shrimp, Fishery Technology, Vol. 47 (1), 2010.

Gilda D., Lio-Po and Celia R. Lavilla–Pitogo. Bacterial exoskeletal lesions of the tiger prawn, *Penaeus monodon*. The Second Asian Fisheries Forum, Asian Fisheries Society, Manila, Philippines, 1990.

Gopakumar, K. and V. N. Pillai. Prawn and prawn fisheries of India, Hindustan Publishing Corporation (INDIA), New Delhi, 2002.

Govindan, T. K. India's foreign exchange earners, Fishery Tech. Vol–8, No. 1, 1971.

Gowda, M. M., M. S. Chandge and M. M. Shirdhankar. Improved shrimp farming technologies, constraints in adoption, Fishing Chimes, Vol. 29, No. 5, Aug, 2009.

Hamner, Willam, M. Krill–untapped bounty from the sea? National Geographic, Vol. 165, No. 5, 1984.

Henrik. B. Nielsen and Pampa Biswas. Shrimp farming in West Bengal, Bay of Bengal News, No. 43, BOBP, September, 1991.

International Finance Corporation (IFC). The Philippine Shrimp Farming Industry, Risks and opportunities for private investors, 1984.

Jha, A. K. and Bijayananda Naik. Microbial biotechnology–an effective means for disease prevention, Fishing Chimes, Vol. 29, No. 3, June, 2009.

Jose, Susheela, K. G. R. Nair, P. T. Mathew, Jose Stephen and P. Madhavan. Modified extensive culture of *P. monodon* using indigenous feed, Fishery Technology, Vol. 39 (1), 2003.

Juario, Jesus, V. Multi-species hatchery, Univ. of the Philippines, 1984.

Juario, V. J. Multi-species hatchery, Univ. of Philippines.

Kakati V. S., K. Y. Telang and C. K. Dinesh. On the fishery of *Acetes johni* of Karwar and Tadri, Marine Information Service, No. 85, CMFRI, Cochin, 1988.

Kasemsarnt, Bung-orn and Arun Rattagool. A survey of the quality of shrimp subsequently used for freezing, Proc. IPFC, 13th Session, Brisbane, Australia, 1970.

Kathirvel, M. and V. Selvaraj. Nursery ground for early juveniles of tiger prawn in Kovalam backwater near, Madras, Marine Information Service, No. 85, CMFRI, Cochin, 1988.

Khor, Stefen. Advance shrimp feed, Fishing Chimes, Vol. 13, No. 9, 1993.

Krishna Menon, M. Notes on the bionomics and fishery of the prawn, *Metapenaeus dobsoni* (Miers) on the south-west coast of India, Indian Journal of Fisheries, Vol. II.

Kurup, B. Madhusoodana. Biofloc in shrimp aquaculture, Fishing Chimes, Vol. 29, No. 1, 2009.

Ling, S. W. Studies on the rearing of larvae and juveniles and culturing of adults of *Macrobrachium rosenbergii* (De Man), Current Affairs Bulletin, No. 35, IPFC, December, 1962.

Macintosh, J. Donald. The status of shrimp farming in South east Asia unpublished paper, 1992.

Manisseri, Mary, K. On the fishery of juveniles of *Penaeus semisulcatus* along the Tinnevelly coast, TamilNadu, Indian Journal of Fisheries, Vol. 29, Nos. 1 and 2, 1982.

Manissery, K. Mary and C. Manimaran. On the fishery of the Indian white prawn *Penaeus indicus* H. Milne Edwards, along the Tinnevelly coast, TamilNadu, Indian Journal of Fisheries, Vol. 28, Nos 1 and 2, 1981.

Mishra, R. N. Shrimps–its postharvest technology and quality control, Proc. National Seminar on shrimp seed production and farming, OUARQT, OSPARC, MPEDA, 1991.

MPEDA,. Overseas markets for diversified product.

MPEDA. Criteria for site selection of prawn hatchery.

MPEDA. Handbook on satellite shrimp farming, 1992.

Mukherjee Kuntal. Marine giant black tiger shrimp (*Penaeus monodon*) importance of aquaculture, problems and immediate remedies, Science and Culture, Vol. 75, No. 5-6, 2009.

Muthu, M. S. and A. Laxminarayana. Induced maturation and spawning of Indian penaeid prawns, Indian Journal of Fisheryes Vol. 24, Nos. 1 and 2, 1977.

Muthu, M. S. Site selection and types of farms for coastal aquaculture of prawns, Proc. First National Symp. on shrimp farming, MPEDA, 1980.

Muthu, M. S., K. A. Narasimham, K. Gopal Krishna and A. K. Sharma. Recent developments in prawn and fish culture in Andhra Pradesh, Marine Information Service, No. 90, CMFRI, Cochin, 1988.

Nair, C. M. and K. R. Salin. Emerging biotechnological approaches in all-male monosex farming of giant freshwater prawn, Fishing Chimes, Vol. 29, No. 5, Aug, 2009.

Nambiar, N. V. and K. Mahadeva Iyer. Common microflora involved in spoilage of canned prawns, Fishery Tech, Vol 7, No 2, 1970.

Namboodri, K. S. Development of an electric shrimp trawl, Reaction of shrimps to low volt direct current, Fishery, Tech. Vol 8, No. 1, 1971.

Nandakumar, G. Observations on the fishery of banana prawn along the north Kanara coast with notes on its schooling bahaviour and migration, Marine Information Service, No. 81, CMFRI, Cochin, 1988.

Nandakumaran, M., D. R. Chaudhuri and V. K. Pillai. Blackening of canned prawn and its preservation, Fishery Tech, Vol 7, No. 2, 1970.

Nielsen, Henrik, B. Shrimp farming in West Bengal, Bay of Bengal News, BOBP publication, Issue No. 43, 1991.

Omori, M. Fishing grounds for spotted shrimp, Yamaha, Fishery Journal, No. 18, 1983.

Patel, B. H. and Ibrahim A. Balapatel. Some observations on the prawn fishery of Gulf of Khambhat, Gujrat Indian Journal of Fisheries, Vol. 29 Nos, 1 and 2, 1982.

Paulraj R. Prawn nutrition and feed development, Proc. Nat. Seminar, Shrimp seed production and farming, OUAT, OSPARC, MPEDA, 1991.

Pillai, S. M. Shrimp hatchery technology in India, Fishing Chimes, Vol. 28, No. 6, Sept, 2008.

Rajan, K. N.; K. K. Sukumaran and Krishna Pillai. On 'Dol' net prawn fishery of Bombay during 1966-76, Indian Journal of Fisheries, Vol. 29, Nos 1 and 2, 1982.

Rajasree, Radhika and B. Madhusoodhana Kurup. Fishery and biology of deep sea prawns landed at the fishing harbours of Kerala, Fishery Technology, Vol. 42 (2), 2005.

Ramamurthy S. G., G. G. Annigeri, and N. S. Kurup. Resource assessment of the penaeid prawn, *Metapenaeus dobsoni* (Miers) along the Mangalore coast, Indian Journal of Fisheries, Vol. 25, Nos. 1 and 2, 1978.

Ramamurthy, S. and M. Manickaraja. An experiment on the culture of *Penaeus indicus* (Milne Edwards) in an estuarine pond at Mangalore, Indian Journal of Fisheries, Vol. 24, Nos 1 and 2, 1977.

Ramamurthy, S. and M. Manickaraja. Relation between tail and total lengths and total carapace lengths for three commercial species of penaeid prawns of India, Indian Journal of Fisheries, Vol. 25, Nos. 1 and 2, 1978.

Rao, C. V. N. Physics in fish processing technology–Fishery Tech. Vol. 8, No 1, 1971.

Rao, P. Vedavysa. Seed requirements for intensive culture of penaeid prawns in coastal waters, particularly in Kerala, Proc. First National Symp. on shrimp farming, MPEDA, 1980.

Rao, R. Mallikarjuna and V. Gopalakrishnan. Identification of juveniles of prawns, *Penaeus monodon* Fabricus and *P. indicus* H. M. Edwards, Proc. IPFC, 13th Session, Brisbane Australia, 1970.

Ravichandran, P. Better management practices in shrimp farming, Fishing Chimes, Vol. 28, No. 7, Oct. 2008.

Ravindranath, K. The Krishna estuarine complex with reference to its shrimp and prawn fishery, Indian Journal of Fisheries, Vol. 29, Nos 1 and 2, 1982.

Sahadevan, P. A good quality feed for shrimp, Fishing Chimes, Vol. 13, No. 10, 1994.

Saisithi, Bung-orn. Fish handling, preservation and processing, SEAFDEC, 1982.

Sarma Kamal, S. Dam Roy P. Krishnan and S. Murugesan. Tiger shrimp brooders of Andaman waters, Fishing Chimes, Vol. 29, No. 10, Jan, 2010.

Schiermeier. Churn, Churn, Churn, Nature, Vol. 447, No. 31, 2007.

Silas, E. G. Status of prawn culture in India and strategy for its future development, Proc. First National, Symp. on shrimp farming, MPEDA, 1980.

Subrahmanyam and K. Janardhana Rao. Observations on the post larval prawns (Penaeid) in the Pulicat Lake with notes on their utilization in capture and culture fisheries Proc. IPFC, 13th session, Bisbane, Australia, 1970.

Subramanian, P, S. Sambasivan, and K. Krishnamurty. A survey of natural communities of juveniles of the penaeid prawns. Proc. First National Symposium on shrimp farming, MPEDA, 1980.

Sudhakara, Rao, G. Prawn fishery by the big trawlers along the north-east coast, Marine Fisheries Information Service, No. 87, CMFRI, Cochin, 1988.

Sukumaran, K. K. Alli. C, Gupta, Uma S. Bhat, D. Nagaraja, H Ramachandra, O. Thippeswamy and Y. Munyappa. Monsoon prawn fishery by "Metabala" along the Mangalore coast, a critical study–Marine Information service, No. 82, CMFRI, Cochin, 1988.

Sukumaran, K. K. and K. N. Rajan. Studies on the fishery and biology of *Parapenaeopsis hardwickii* (Miers) from Bombay area, Indian Journal of Fisheries, Vol. 28, Nos. 1 and 2, 1981.

Surendran, P. K. and K. Mahadeva Iyer. Antibiotics in the preservation of prawn, Fishery Technology, vol. 8, No I, 1971.

Suseelan, C. and M. Kathirvel. A study on the prawns of Ashtamudi back waters in Kerala, with special reference to penaeids, Indian Journal of Fisheries, Vol. 29, Nos. 1 and 2, 1982.

Swaminathan, M.S. Shrimp farming–A new dimension to the scientific utilization of our aquatic wealth, Proc. First National Symp. on shrimp farming, MPEDA, 1980.

Thomas, M. M. A new record of *Epipenaeon ingens* Nobili (Bopyridae, Isopod) parasitic on *Penaeus semisulcatus* from Palk Bay and Gulf of Mannar, Indian Journal of Fisheris, Vol. 24, Nos 1 and 2, 1977.

Thomas, M. M. Decapod crustaceans new to Andaman and Nicobar Islands, Indian Journal of Fisheries, Vol. 24, Nos 1 and 2, 1977.

Tim Bostock. Better feeds for small-scale, shrimp farmers, Bay of Bengal News No. 42, BOBP, June, 1991.

Verghese, P. U. Potentials of brackish water prawn culture in India, Proc. First National Symp. on shrimp farming, MPEDA, 1980.

Yousuf Haroon, A. K. Freshwater prawn farming trials in Bangladesh, Naga, ICLARM July, 1990.

Yunker, M. Tips for buying and stocking healthy fry, Iloilo, Philippines.

Index

P

Paddy and prawn production, 268

Palaemon, 6, 7

Pandalidae, 22, 51

Pandalopsis, 24

Pandalus, 23, 24, 25, 26, 27

Parapandalus, 11

Parapenaeopsis, 8, 16, 17, 36, 49, 65

Parapenaeus, 18, 53

Penaeid prawn, 2, 39, 40, 67

Penaeopsis, 18, 53

Penaeus, 7, 8, 18, 19, 20, 35, 37, 38, 39, 48, 63, 64

Plesionika, 11

Porto Novo, 125, 146

Post larva (PL), 84

Postharvest handling, 269, 304

Post-larva, 78, 92, 116, 117

Prawn farming in India, 263, 292

Prawn farming in S.E. Asia, 222, 223

Prawn hatchery, 227

Prawn processing, 312, 313, 314

Preservation, 191, 193

Pretreatment, 199

Probiotics, 291

Production, 43, 44, 68

Pulicat Lake, 93

Q

Quality control, 198, 206

Quality prawn fry, 252

Reaction of prawns, 189

Rearing in floating cages, 262

Remedial measures, 259

Resources, 71

Responsible prawn farming, 287

S

Sasson Dock, 151, 150, 153

Scoop net, 111

Seed production, 226

Seed resources, 240

Semi-dried prawn, 207

Shore seine, 112

Shrimp hatcheries, 216

Sicyonia, 20, 53

Soil characteristics, 293

Solenocera, 20, 21, 53

Spawners, 280

Spawning, 83, 115

Spoilage, 191

Sri Lanka, 55

Suitable sites, 216

T

Tamil Nadu, 60, 10, 135, 136

Taxonomic characters, 3, 4

Testis, 112

Thailand, 56

Thawing, 202